Negotiating the Past

Negotiating the Past

The Making of Canada's National Historic Parks and Sites

C.J. TAYLOR

McGill-Queen's University Press
Montreal & Kingston · London · Buffalo

ISBN 0-7735-0713-2

Legal deposit first quarter 1990
Bibliothèque nationale du Québec

Printed in Canada on acid-free paper

This book has been published with the help of a grant from the Social Science Federation of Canada, using funds provided by the Social Sciences and Humanities Research Council of Canada.

Canadian Cataloguing in Publication Data

Taylor, C. James, 1947–
Negotiating the past: the making of Canada's National Historic Parks and Sites
Includes bibliographical references.
ISBN 0-7735-0713-2
1. National parks and reserves—Canada—History.
2. National parks and reserves—Government policy—Canada. 3. Historic sites—Canada—History. 4. Historic sites—Government policy—Canada. I. Title.
FC215.T39 1990 333.78'3'0971 C89-090469-3
F1011.T39 1990

For RWT: 1919–73

Contents

Preface

The people and places that comprise the national historic parks and sites program form one of the country's largest cultural institutions. Certainly no other federal agency spends so much to tell so many people about their past. Each year thousands of tourists visit Fortress of Louisbourg National Historic Park where they experience first hand life in an eighteenth-century town and learn about the fortifications that made Louisbourg so important to the defence of New France. Similarly, visitors to Lower Fort Garry see what life was like in a nineteeth-century fur trading post and at Fort Rodd Hill National Historic Park they learn about a piece of the British imperial defence system. At other historic parks visitors see aspects of domestic life from a number of periods and regions. And many travellers have come across the familiar maroon markers designating sites of national historic interest. Battlefields, explorers' landfalls and homes of the rich and famous have all been commemorated as national historic sites. Having operated since 1919, the program has grown into a considerable enterprise. There are now over seven hundred national historic sites and more than forty national historic parks distributed across the country with facilities for visitor reception, artifact conservation, and historical interpretation. The program is large in other ways. It employs scores of specialists conducting archaeological and historical investigation, collecting artifacts, and designing exhibits. It has become a prominent player in the preservation of what is trendily known as the "built heritage." Besides developing high standards for its own restoration and preservation projects, it has been a leading figure in co-operative endeavours to preserve historic districts in Halifax, Québec City, and Niagara-on-the-Lake. Behind all this there is a history that is itself worthy of commemoration for its national significance.

With an extensive system of sites and a large and active administration, it is surprising that the national historic parks and sites program has not attracted more attention than it has. Partly because its administration is buried in the larger national parks operation which has a strong and distinct identity of its own, partly because its operations are so decentralized, the national historic parks and sites program has a comparatively low profile as a federal agency. Yet its varied activities across the country make its history fascinating in a number of respects. Because national historic sites stem from subjective views of national development, they assume significance as cultural artifacts beyond the historic events they commemorate. Collectively, regionally, and locally, national historic sites are important sources of information about our national outlook. This realization leads to a series of questions: what constitutes national significance; why are particular sites included in the system and others not; what is the relation between the government and national identity; and what does all of this tell us about ourselves as a country? These questions can only be answered by a comprehensive history of the agency.

The history of national historic parks and sites sheds light on the sociology of the federal government. The interaction of the program with various participants shows the multifaceted relationship of the federal government both internally and to the society with which it interacts. Although the heritage program was and remains a very minor concern of the government as a whole, its history illustrates the way in which the federal system works in trying to bring regional aspirations and identities into a federal forum. Instead of using history to exercise hegemony over dissident outlooks, the state encouraged regional interests. We see, for example, the case of Fort Chambly in Québec, a largely British fort interpreted as a centre of French-Canadian folk culture.

The history of historic sites also reveals the layers of circumstance which impinge on a federal cultural program. We see by example how policy is established, how decisions are made, and how government initiatives are both encouraged and thwarted. Decisions affecting spending on heritage projects, for instance, often had less to do with the particular merits of the case than with larger fiscal and political concerns. So, the success or failure of the program often did not depend on the capacity of the administration to carry out its responsibility, but on its ability to respond to changing tides of government priorities and the shifting winds of public opinion.

With the history of such an interesting agency waiting to be written, I was drawn into the topic by my special interest in the program. Like many people, I was first drawn to historic sites by the maroon plaques that dot the countryside. My involvement in them was casual, however, until I joined the organization responsible for these sites, then called Parks Canada, some fifteen years ago. In the course of writing and rewriting plaque inscriptions, I became intrigued about the history of particular sites: why had they been selected in the first place, and why have attitudes toward them changed over time? These questions led me to wonder about the organization that looked after national historic parks and sites: why was it established and how had it evolved to meet its objectives? In this way I used my historical training to explain the context of my work. As I looked into these questions and talked them over with friends and colleagues, I came to realize that the answers could be interesting to others concerned with heritage, government programs, historiography, or generally Canada's past. It was to marry these personal and public interests that the present book was conceived.

In writing this history I was faced with a number of practical and theoretical problems. Practically, my employer had little interest in official navel gazing, yet a study of this scope required considerable time and resources. Fortunately, the bureau was willing to partly fund a period of educational leave and I wrote the core of this study as a doctoral thesis for Carleton University. At Carleton, a doctoral fellowship from the Social Sciences and Humanities Research Council made student life a bit more comfortable. Back at parks headquarters, the bureau kindly assisted with word processing.

Such a broadly dispersed topic as the history of historic sites posed a number of problems for the lone researcher working in an underdeveloped field. The help and encouragement from friends, colleagues and teachers considerably eased the way. Individuals associated with the national historic parks and sites program were generous in encouraging my work and sharing their time and research: Margaret Archibald, Phil Goldring, David Lee, Julian Smith, Terry Smythe, and Janet Wright at headquarters; Rick Stewart in Calgary; Barbara Schmeisser in Halifax; and John Johnson at Louisbourg. J.M.S. Careless, former chairman of the Historic Sites and Monuments Board of Canada, and David E. Smith, Saskatchewan member of the board, were supportive, while Professor Smith read the present manuscript in draft form and made helpful suggestions. Ernest Côté, assistant deputy minister and then deputy minister responsible for the program during the 1950s and 1960s, provided a rare inside perspective of senior management during that period.

At Carleton University, Blair Neatby proved to be everything a student would want in a professor. He was immensely helpful in apprising me about the political context of the period of the study and teaching me the intricacies of forming a coherent narrative from a mass of material. Subsequently, while I was making the thesis into a book, he kindly shared his insights gained from researching his history of postsecondary education in the postwar era. Hal Kalman urged me to continue and let me have the run of his superb library on preservation history and practice. Janet Wright, a friend as well as a colleague, encouraged me at every stage and shared her insights gained from her work on the history of the Chief Architect's Branch of the Department of Public Works. And finally, Colleen Gray, who edited the manuscript for McGill-Queen's, was extremely helpful in rooting out inaccuracies, inconsistencies, and sloppy writing. Needless to say, the responsibility for the content and presentation of this book rests entirely with myself. Moreover, this book is in no way an official history, and the views expressed are those of the author, not those of the Canadian Parks Service.

Despite all this help and encouragement, there were still theoretical problems that I had to solve by myself. The program is both large and unfocused. As complex as the program itself is the larger context of both the national parks program and the Canadian heritage movement. Yet there are few useful studies that could usefully define my work. The first difficulty, then, was to overcome the problem experienced by many pioneers – to orient myself to a new field.

There are several ways that I could have approached this history. I could have regarded the federal program as just one aspect of a larger heritage movement. Such an approach would have chronicled every detail of every initiative to preserve historic sites much like Charles Hosmer's massive history of the American preservation movement, *Presence of the Past* and *Preservation Comes of Age.*[1] Or I could have assumed a set of standards for national preservation and judged the history of the federal program accordingly. Neither of these extremes seemed particularly satisfactory. The first was unsuitable because I did not wish to write merely a descriptive compendium of heritage preservation in Canada; the second, because it judges the past from the perspective of the present. Neither approach would have adequately explained the dynamic process behind a federal agency, and I wished to write a history that explained a government program as much as it documented efforts to commemorate and preserve the past.

The approach I chose was to interpret the development of the federal historic sites program according to a dynamic process I term "the politics of historic sites." A cultural program like that of national historic parks and sites involves a number of relationships held together in this dynamic process: the relation between the state and society, regional and national perspectives, and history and national identity all interact in this dynamic process that becomes manifest in the selection, interpretation, and development of historic sites. I call this the politics of historic sites because the parks and sites are the result of varied and sometimes conflicting concerns that are worked out in an ongoing process of negotiation.

This process really lends itself to treatment by a narrative history. The politics of historic sites can truly only be appreciated through its history, for what Ortega y Gasset said of mankind is equally true of the historic sites program: it has no nature, what it has is history.[2] At any point in time the program is the product of past decisions and future considerations, and these can best be understood as part of an ongoing process. The development of Louisbourg, for example, has undergone a number of radically different phases and may yet be transformed again. Each of these phases was shaped by a nexus of personalities, events, and attitudes playing off one another. Similarly the organization's growth was affected not by a predetermined plan but by a complex of interrelationships regulated alternatively by inertia and activity. In the end the system is the product of its own past. This history can chronicle the events that led to the selection and development of particular sites. It can also recreate some of the dynamic relationships which form the politics of historic sites.

The politics of historic sites involves a number of key players. As the participants in this history are many and various, it may be helpful to introduce the main players. A major participant is something that can be loosely defined as the national heritage movement, comprised of local groups and national voluntary organizations that promote the identification and treatment of historic sites as national shrines. Pressure from this movement led Ottawa to acquire individual sites and continues to shape the direction of the heritage program. Yet the movement pulls the program in various ways as local groups present different interpretations of national history. These competing perspectives were revealed even before the inauguration of the historic sites program. In 1908, for example, the federal government was persuaded to preserve the Plains of Abraham as a national battlefield. Although there was consensus over the development of the site as a national treasure, opinion differed as to what it signified. Anglophone Canada saw the plains as one of the great scenes from British imperial history, the battlefield where Wolfe vanquished Montcalm. Francophone Canada, on the other hand, wished to infuse the site with more vague historical associations, linking it to Champlain's founding of Québec. Both these groups saw the plains as a central symbol of their view of the nation's historical development, yet these historical interpretations were as different as their views of contemporary Canada. Ultimately, the government had to accommodate both of these perspectives into the presentation of the site and the result is a rather bland park with little didactic meaning. Different views of Canada lead to different views of national history. Questions such as who discovered Canada, who were the 'good guys' in the French-English colonial wars in North America, what was the significance of the Riel rebellion evoke different responses from different groups and regions.

There were other divisions among the participants of the heritage movement concerning the selection and treatment of sites besides those stemming from the interpretation of national significance. There was another group more concerned with preserving examples of the historical landscape. While those who wished to use historic sites as symbols of national history were more interested in advertising their significance as part of a national history through commemorative markers, the other group was more concerned with preserving historic sites as national treasures. Local advocates for the development of the ruined fortress at Louisbourg, for example, were not too concerned with whether it was seen as the site of a great French fortress or a great British victory so long as it was developed as a heritage attraction. Among the preservationists there are other questions, conflicting views of treatment as well as selection: should a ruin be preserved as it was, as a romantic reminder of a vaguely apprehended past, or should it be developed into a more popular attraction, made more useful for ends of tourism and education?

With so many forces behind the movement to create national historic parks and sites, it was the responsibility of the government to resolve the conflicting concerns and perspectives into a single program. As the final arbiter, the government had to decide what was to be included in the system and, *ipso facto*, defined as having national historic significance. This raises the possibility that the government imposed its own objective criteria and so dominated public opinion in some hegemonic way. A quick glance, however, at historic site designations suggests that this is not the case. National historic sites reflect regional identities. Sites commemorating the Loyalists and the War of 1812 are predominantly in Ontario. Québec sites favour events from the French régime while the west includes a large proportion of sites commemorating exploration and the fur trade. The government reflects, it does not shape, the values of a geographically diffused and pluralistic society. In implementing a heritage program, it internalized the many and varied attitudes toward the past held by Canadian society.

The interaction between the government and the various concerns of the heritage movement introduces another set of relationships, one that is internal to the program. When we try to identify those in charge of the federal heritage undertaking we discover, as with other government programs, a welter of competing jurisdictions. In the case of the historic sites program, there are three main divisions – the executive, the civil service, and the advisory body – each of which includes smaller units of often competing and contradictory concerns. Each brings different priorities to the program – economic and cultural development, conservation of physical resources, and commemoration of a national heritage – which together influence its direction.

Strictly speaking, the government is the executive, that is, the cabinet, which is made up of ministers of the Crown. Its task is to administer broad government policy, but this is influenced by the party system so that members of Parliament, who formally constitute the legislature, can bring regional and local concerns into the process of executive decision making. Both larger government policies and local concerns tend to affect the government's relationship with its heritage program. During the 1920s, for example, fiscal restraint severely inhibited the early development of historic parks, while a middepression policy decision to aid economic recovery through public works brought about a dramatic change in the historic sites program. Similarly, government decisions taken in the late 1950s and early 1960s to aid regional economic expansion influenced historic park development at Louisbourg and Dawson. While the executive no doubt adhered to idealistic principles such as national unity in promoting national historic sites, it also viewed heritage developments as tourist attractions providing economic benefits to regions where votes were important.

Subordinate to the executive, but nevertheless an important agent in implementing the historic sites program is the administration of the national parks service. Made up of experts and career civil servants, it provides the detailed knowledge upon which decision making depends. It is closely tied to the executive through the minister who has final responsibility for its actions. But with the mandate to manage the day to day affairs of an array of ongoing government programs, the parks service has considerable autonomy and possesses sufficient independence to have its own stated policies regarding these programs. It is further removed from government control by virtue of the fact that it forms merely a part of a department. Until 1936 the national parks service was buried in the massive Department of the Interior. Subsequent to that it formed a minor branch in a succession of departments from Mines and Resources to Indian and Northern Affairs. It now forms part of the Department of the Environment.

Within the administration are a number of distinct entities. The assistant deputy minister is responsible for the entire parks program. He or she channels the views of the executive and the minister to the branch, imposes his or her own views on the administration, and mediates the competing perspectives of subordinate divisions. Until the mid-1950s there was a director of national parks who supervised a number of units such as natural parks, historic parks, and engineering services. Subsequently, the historic sites program received greater autonomy, although it still remains part of the larger national parks administration and is influenced by the larger priorities of that organization. Charged with the preservation and development of natural landscape, the national parks service tends to regard historic parks and sites as an extension of this mandate. The development of historic parks for recreational use reflects this larger branch concern. The bureau specifically charged with looking after the historic sites program has to compete for money and personnel with other units of the parks branch. When historic sites were a low priority, it was difficult to obtain necessary resources to run a basic service. Yet when historic sites assumed a higher profile in the branch, it was in danger of being taken over by rival units such as engineering services.

The Historic Sites and Monuments Board of Canada is the third agent in the federal historic sites program. Formally appointed to advise the minister concerned on the designation and development of national historic parks and sites, it in fact works closer with the parks administration which is charged with implementing its recommendations. The board interacts with the executive and the administration in defining the program, trying to accommodate itself to the circumstances dictated by its larger partners. Because it has minimal influence on public spending, the board has little to do with site development, concentrating instead on site selection. More than any of the other players in the program, then, the Historic Sites and Monuments Board is most concerned with the question of the national significance of sites and its relation to regional perspectives.

The board itself is not a homogeneous entity as its members are drawn from regional constituencies. So, while it is part of a cultural agency endeavouring to apply national criteria to the selection of historic sites, it introduces regional and provincial perspectives through the representation of its members. At the same time some of the board's members have acted as if they were custodians of a national culture and have consciously striven to present ideals of nation building and civilization to what they considered to be a disparate and aimless society. In this way competing views about the nature of a national history interact in a single forum, and the board is closely tied to the relationship between national identity and regional self-expression.

The cluster of relationships affecting the history of the historic sites program operates from both without and within. On the one hand, there are competing regional and local concerns for the commemoration, preservation, and development of local shrines as national monuments. On the other hand, there are the different strands of the program itself: the executive, the administration, and the advisory board. All of these bear on the ultimate treatment of any historic park or site.

Competing regional concerns and the fractional nature of the state make implementing a heritage program extremely difficult. The task is further complicated by the physical nature of historic parks and sites. Unlike museum artifacts or paintings, historic sites remain rooted to their regional settings. Control over them, therefore, remains costly and is subject to continual negotiation with local concerns. Even when a property is formally under the department's jurisdiction as a national historic park, its development is subject to local pressures. The treatment of Louisbourg and Fort Chambly, for example, incorporate both local and government aspirations. Places selected as national historic sites remain constantly on display. While they can be re-interpreted to reflect contemporary historical opinion, the scope of their interpretation is limited by the historical associations of the site. Thus the system remains encumbered by a large number of military sites designated at a time when military defence had greater importance in Canadian historiography than it does today.

The diversity of opinion on national heritage, the fragmented nature of state administration, and the cumbersome physical nature of historic sites inhibit the development of a rational plan or policy for the selection, preservation, and interpretation of heritage places.

With all of these competing concerns driving the politics of historic sites, it is difficult to articulate and maintain coherent objectives for the program. Is the program meant to preserve a network of national shrines or develop cultural resources for educational or economic ends? Is it to impose larger ideas of national unity on a disparate population, or represent regional perspectives in a national pantheon of historic sites? These difficulties are borne out in the history of the program. As competing aims came into play the program was often badly out of focus despite repeated attempts to direct it toward well-defined objectives.

In presenting the history of the federal program in this way I do not mean to ignore the importance of ideas. Ideas about treating historic sites that flourished in international and national heritage circles certainly influenced attitudes within the federal program. James Harkin, the first commissioner of national parks, was greatly influenced by John Ruskin's dictum that reconstruction (as opposed to simple preservation) was destructive. The decision to reconstruct colonial Williamsburg in the United States in the late 1920s certainly affected the approach to the treatment of the Fortress of Louisbourg. Ideas about the treatment of historic sites have also shaped the following narrative. I have assumed, for instance, a separation between the activities of commemoration and preservation which, while not always distinct, explains differences in emphasis in the work of the Historic Sites and Monuments Board and the Canadian Parks Service. This assumption is not shared by the federal agency today, which maintains that preservation is merely a function of commemoration. Still, I prefer my distinction as it better defines the attitudes of these two groups. I chose not to elaborate on these theoretical principles, however, except where preservation ideology impinges on federal policy. Ideas alone were nothing without the participants to express them, and it is with these players that we are primarily concerned.

The revitalized Historic Sites and Monuments Board of 1923. (National Archives of Canada [NA], PA 66730)

Fort Chambly, Québec, in the 1870s. A natural historic park. (Notman Photographic Archives, McCord Museum of Canadian History)

John Clarence Webster (1863–1950. As New Brunswick member of the Historic Sites and Monuments Board from 1923–50, he was responsible for many program initiatives in the Maritimes. He served as chairman from 1945–50. (NA, PA 66731)

Fort Anne, Annapolis Royal, Nova Scotia. Established as a national park in 1917, it was, with its fortifications and landscaped grounds, the archetypical national historic park of the prewar period. (Canadian Parks Service)

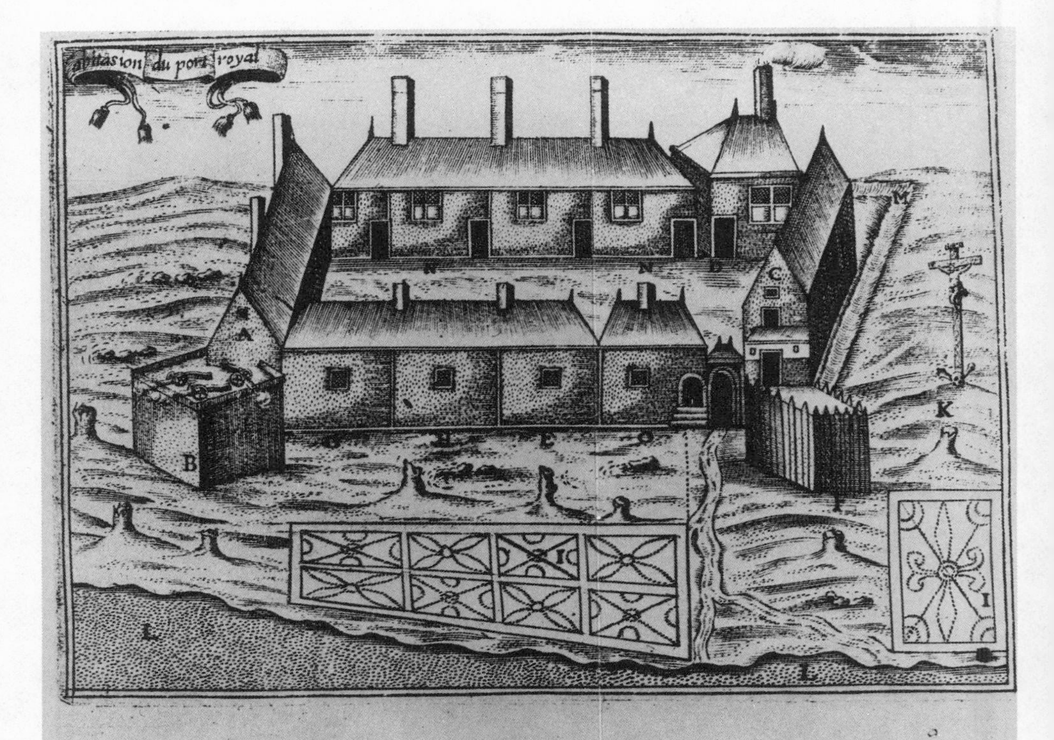

Champlain's picture plan of the settlement of Port Royal ("abitasion du port royal"), reprinted in H.P. Biggar, ed., *The Works of Samuel de Champlain* (Toronto: Champlain Society 1922), 1:372–3. Following the First World War, antiquarian and scholarly research uncovered numerous records relating to historic buildings. This information encouraged efforts in the 1930s to reconstruct replicas such as the Habitation Port Royal and, later, the Fortress of Louisbourg.

Fort Malden, Amherstburg, Ontario. This property was acquired in the 1930s and 1940s as a result of the lobbying of a local heritage group and against the advice of the Historic Sites and Monuments Board. Nonetheless, it conforms to the archetype of the prewar national historic park. (Canadian Parks Service)

Prince of Wales's Fort, Churchill, Manitoba. Aerial photograph taken in 1947. One of the first historic properties to be acquired by the program, it was restored in the 1930s and 1950s. (Canadian Parks Service)

Restoring Prince of Wales's Fort in the 1950s. (Canadian Parks Service)

Fort Langley, Langley, BC. Acquired through the efforts of local heritage groups and the BC member of the board, W.N. Sage, Fort Langley was reconstructed in the 1950s. (Canadian Parks Service)

Fred Landon (1880–1969). Ontario member of the Historic Sites and Monuments Board from 1932–58. As chairman from 1950–8, he helped lead the board into a new era. (NA, Royal Society of Canada Portraits, Acc. #1943–23)

E.A. Côté (1913–). As assistant deputy minister of the Department of Northern Affairs and Natural Resources from 1955–63, he was the first senior civil servant since Harkin to take an active part in the historic sites program, introducing new approaches to the protection of historic architecture. He was deputy minister from 1963–8, when the program became involved with megaprojects such as the reconstruction of the Fortress of Louisbourg. (Department of External Affairs)

Former Grey Nuns' convent, now the St Boniface Museum, St Boniface, Manitoba. Its restoration in the 1960s was the result of one of the first co-operative agreements negotiated by the program where the federal government aided local initiatives to preserve historic buildings. (Canadian Parks Service)

Palace Grand Theatre, Dawson, YT. This replica was built in 1962 after it was found that preserving the original as an operating theatre would be too difficult and costly. The project initiated the massive heritage development at Dawson. (Canadian Parks Service)

Above and preceding page: The reconstruction of the fortress of Louisbourg in the 1960s was the greatest single heritage project undertaken by the parks branch. It set the standard for national historic parks for the latter part of this century. (Canadian Parks Service)

Negotiating the Past

CHAPTER ONE

Legacy, 1887–1919

Although the federal program for creating and caring for officially sanctioned national historic parks and sites was not inaugurated until 1919, its roots went back to the previous century when Ottawa first became involved in recognizing the young nation's heritage. In tracing these roots, we see the emergence of the principal participants who would influence the direction of the federal heritage program between the two world wars: the national heritage movement, including its regional and national components; individuals such as Ernest Cruikshank, James Coyne, and Benjamin Sulte; and the national parks branch of the Department of the Interior headed by the indefatigable James Harkin. This formative period also witnessed the identification of what would become key properties in the national historic parks system.

THE FORMATION OF A NATIONAL HERITAGE MOVEMENT 1887–1919

The heritage movement in Canada was largely the manifestation of the work of local organizations which promoted their sites as national resources and coalesced at a national level through umbrella agencies such as the Royal Society of Canada and the Historic Landmarks Association. In some respects these groups were remarkably disparate, including patriotic societies such as local chapters of the Imperial Order Daughters of the Empire and the St Jean Baptiste Society, regular historical societies, and more broadly based cultural organizations. Rooted in local contexts, they were particularly prone to regional biases. Yet they also shared some remarkable similarities. They tended to be dominated by the establishment – members of the local élite and well-educated members of the propertied class – and were often concerned with promoting cultural ideals, ideals which looked beyond local differences and sought national cohesion.

In Ontario, history and nationalist ideology were joined through a cluster of beliefs generally referred to as the loyalist tradition. One of the greatest exponents of this tradition in the writing of Canadian history at the turn of the century was George Taylor Denison. His views on the importance of Loyalists and their ancestors in the formation of a distinctive Canadian society were expounded in his 1904 presidential address to the Royal Society of Canada. His paper, "The United Empire Loyalists and their Influence upon the History of this Continent," began by describing the Loyalists as the respectable element of American society driven north by lawless agitators revolting against the British Crown.[1] In British North America they provided a firm basis on which Canada could emerge as an independent state in North America loyal to the Crown. Denison described them as a homogeneous mass, possessing varying degrees of wealth, but tied together by their doctrine of "Fear God and Honour the King."[2] Thanks to their influence Canadian society was distinguished by being more law abiding than the United States for "the pious, God-fearing men who had made such sacrifices for their principles, were a community almost free from crime."[3] More importantly, they had consistently led the defense of Canada against American aggression. Loyalists and their children saved the day for Upper Canada during the War of 1812, taking a leading part in the local militia units. Again loyalist descendants had been quick to come out in support of the civil authority against the republican rebels in the 1837 rebellion. Similarly, they had staunchly defended the Crown during the Fenian raids and the Trent affair. And, more recently, they had led the fight against the insidious movement for commercial union with the United States. To Denison, the Loyalists, in sum, embodied a distinctive heroic tradition which all Canadians could revere. "Canadians may well be proud of the founders of this country, and all classes should combine to perpetuate the principles which have guided us so well in the past."[4]

More recently, Carl Berger elaborated on this notion of a loyalist tradition to show how history was employed by a political élite in the late nineteenth century to justify its larger view of imperial unity. "Not all historical work during these years derived from the loyalist cult. But history was the chief vehicle on which the loyalist tradition depended for its credibility upon the assumption that the past contained principles to which the present must adhere if the continuity of national life was to be preserved."[5] This was a defensive form of nationalism, based on a fear of absorption by the United States. It inspired a faction of contemporary politics which was militantly against free trade with the United States with examples from the past of patriotic stands against the Yankee invader. The importance of the past to the conservative élite in Ontario meant that history was promoted with missionary zeal. As a result, interest in history transcended partisan lines. Gerald Killan, in his history of the Ontario Historical Society, has noted that membership in local historical societies had an extremely broad base, including more than followers of the conservative élite.[6] Nevertheless, he also points to the popularity of the loyalist tradition: "nineteenth century Ontarians embraced history to promote a variety of causes, the chief of which was to cultivate a British Canadian nationalism."[7]

This nationalist ideology was used to gain federal support for a number of Ontario historic sites. A local heritage activist who followed this route was Canon George Bull who helped organize the Lundy's Lane Historical Society in 1887 with the immediate aim of developing Lundy's Lane Battlefield, but also to help spread the loyalist doctrine. The Battle of Lundy's Lane occured in 1814 when a large American force attempted to invade Canada from across the Niagara River. British regulars and Canadian militia units under the command of General Drummond repulsed the Americans after some bitter hand to hand combat.[8] The Canadians experienced a decisive victory, but the Americans managed to inflict heavy casualties and then retreat in order. Although a more significant victory had taken place the previous year when Canadian Indians surrounded and captured some three hundred Americans at the Battle of Beaver Dams, Lundy's Lane was significant for the honour it gave the Canadian militia, including the York volunteers who had fought with distinction. Hence it was a prominent event in the lore of the loyalist and the militia myth.[9]

Canon Bull wished to remove the remains of the Canadian dead from the battlefield to a large common grave and mark it with a fitting monument. Although initial attempts to raise sufficient funds by public subscription failed, the federal government was persuaded to lend a hand to complete the project. Ottawa commissioned E.E. Taché to design a monument that would be erected on War of 1812 battlefields and, in 1895, granite pillars forty feet high were unveiled at Châteauguay, Crysler's Farm, Lundy's Lane, and, later, at Stoney Creek.[10] The government contributed five thousand dollars to the building of each of these, which was disbursed through the Department of Militia and Defence. Although the department oversaw the spending of the money, local historical societies were still actively involved, and in the case of the Lundy's Lane monument, the site became the responsibility of the Queen Victoria Niagara Parks Commission, an agency of the provincial government founded in 1893. The inscription for the monument represents the view of history promoted by Canadian imperialists and was likely composed by Canon Bull or George Taylor Denison.[11] It reads:

Erected by the Canadian Parliament in Honour of the Victory Gained by the British Canadian Forces on this Field on the 25th of July 1814, And in Grateful remembrance of the Brave Men who died that Day fighting for the Unity of the Empire.
1895

Although the emphasis on imperial unity is peculiar to monuments of this region, the Lundy's Lane Battlefield is in many ways an archetypal national historic site, and the implication of the interpretation – that the event affected the future development and outlook of the nation – is common to other sites commemorated at the turn of the century.

But the imperialists' view of history did not go unchallenged in Ontario. Tied strongly to the Tory ideology of no truck nor trade with the Yankee, it was natural that Liberal opinion would try to modify it, especially at this time, when the Reciprocity Treaty was being hotly debated. One of the chief proponents of this revisionist ideology was Goldwin Smith, who argued for the development of Canada as an independent state in harmonious co-operation with its continental neighbour. Smith was a former Oxford University professor who enjoyed considerable prestige in Canada, particularly in Toronto intellectual circles. Consequently, when he argued that the Lundy's Lane memorial should honour the dead of both sides, not just the loyal Canadians, he posed a considerable threat to Bull and his acolytes. Goldwin Smith's proposal was supported by the Liberal Toronto *Globe* in an editorial of July 1895 which noted in part: "Now, much as we cherish our dead, revere the memory of Lundy's Lane and rejoice in our Canadianism, was there anything discreditable in Mr. Goldwin Smith's actions? Was his an ignoble thought? On the heights of Abraham the Monument to Wolfe and Montcalm stands, a lesson in reconciliation to all the world. Is it a vain hope that Lundy's Lane may read the same lesson to generations that follow?"[12] But this proposal would have found little support with the presiding Conservatives at Ottawa. It smacked too much of Liberal continentalism.

Despite the influence of revisionists like Goldwin Smith, there were a number of patriotic and historical societies, like the Lundy's Lane Historical Society, which were formed to promote specific heritage issues. At Hamilton, for instance, an organization dominated by women members of the local élite brought about the purchase of the Gage homestead. The property was important as the site of the Battle of Stoney Creek in the War 1812, but also included a pioneer building which was to be restored by the women's committee.[13] Elsewhere in southern Ontario there were a number of nonstrategic military sites administered by the Department of Militia and Defence which became heritage issues as local historical societies adopted them as preservation causes. Many of these forts dated back to the War of 1812 and some had subsequently seen action during the 1837–8 rebellions and the Fenian raids. By the time the British army left them in 1870 they were quite useless as defensive establishments, yet their history made them shrines for local historical and patriotic groups. When, therefore, early in this century, the government took steps to divest itself of some of this real estate, it met strong opposition from local societies and members of Parliament. Typical of these properties was Fort Malden in Amherstburg.

Fort Malden had been built by the British army during the War of 1812 and was later garrisoned by local militia units. In 1875, following its transfer to the Canadian government, the property was put up for sale. Many local citizens still considered the property to be a government responsibility, and in 1904 the residents of Amherstburg petitioned the minister of Militia and Defence to re-acquire the property as a national historic park.[14] No definite action followed this petition, and in 1909, when the land was threatened with subdivision, the newly formed Amherstburg chapter of the Imperial Order Daughters of the Empire petitioned the governor-general to nationalize the property.[15] Again in 1912, the centenary of the fort, pressure was put on the government to acquire the site, and a delegation met with Prime Minister Borden. Although Borden was sympathetic, further discussions were interrupted by the outbreak of war.

While many sites such as Lundy's Lane and the Stoney Creek Battlefield were promoted for ideological ends, there was also concern for the preservation of an historic landscape. At the turn of the century the city of Hamilton purchased the home of Sir

Allan MacNab, a palatial residence called Dundurn Castle, splendidly situated on park-like grounds overlooking Lake Ontario. The plan was to turn the building into "a museum for the preservation of historic relics and scientific specimens, and the exhibition of art treasures."[16] Heritage groups sponsored historic sites for both patriotic and aesthetic reasons. At the opening of Dundurn Park noted, Canadian historian, J.A. Bourinot, pointed to the meaning which history gives to the landscape. "The various human forces that have exercised such potent influence on the development of Canada, have at one time and another met on this historic height, or by the side of the beauteous bay below."[17]

Other organizations promoted an interest in the past for scientific reasons. The Canadian Institute was a small organization of academics, professionals, and others who met regularly in Toronto to discuss scientific subjects. History and archaeology were included in their curriculum, and one of their members, David Boyle, undertook to form a small museum at the institute's headquarters. In 1887 he was given a small salary through the provincial ministry of education and assumed quasi-official status as provincial archaeologist.[18] He undertook a number of investigations of prehistoric sites across southern Ontario and preserved and catalogued hundreds of artifacts for the institute's museum. His approach to the past was avowedly empirical. He argued that primitive societies should be studied scientifically and systematically not to demonstrate the superiority of contemporary society but to discover empirical truths. "The myths and superstitions of primitive folk," he wrote in his first annual report, "their social organization, their germs of constitutional government, their daily occupations, their farms, ceremonies, games and amusements, the mechanical methods and devices they employed, and the examples of their handicraft – all these must ever possess an increasing interest to thoughtful persons generally, but more especially to those whose desire it is to study civilization 'in its wide ethnographic sense.'"[19] Boyle helped to popularize a widespread interest in Ontario prehistory and drew attention to a number of important archaeological sites.

One of the best known of these sites was the Southwold Earthworks near St Thomas, Ontario. It became the concern of the Elgin Historical and Scientific Institute, whose members viewed it not only as an important scientific subject, but also as a local landmark. Nobody was certain of the history of these mounds of earth, although it was correctly assumed that they were the remains of a prehistoric (that is, precontact) fortified village. James Coyne, local historian and president of the institute, wrote in 1895: "In the part of the Township of Southwold included in the peninsula between Talbot Creek and the most westerly bend of Kettle Creek there were until a comparatively recent date several Indian earthworks, which were well-known to the pioneers of the Talbot Settlement. What the tooth of time had spared for more than two centuries yielded however to the settler's plough and harrow, and but one or two reminders of an almost forgotten race remain to gratify the curiosity of the archaeologist or the historian."[20] Coyne's remarks introduce yet another motivation for the heritage movement – the concern for a landscape that was being rapidly transformed by settlement and other aspects of the modern world.

After 1899 the efforts of local heritage groups were co-ordinated at the provincial level by the Ontario Historical Society. The society became involved in a major preservation issue in 1905 when it confronted the city of Toronto to save "Old" Fort York. The fort, situated on the grounds of the Canadian National Exhibition, was threatened with destruction to make way for a street railway line, but the historical society, in conjunction with local preservation groups, successfully battled various levels of government to have the structure preserved. During the struggle, the society formed an Historic Sites and Monuments Committee to co-ordinate "the save Fort York campaign" and subsequently came up with a development plan for the site. Later, this committee also promoted other preservation issues.[21]

Québec had its own nationalist perspectives which influenced the selection of historic sites in that province. Jacques Monet, in *The Last Cannon Shot*, has described one strain of French-Canadian nationalism emerging in the midnineteenth century. This is epitomized by the words of Sir Étienne-Paschal Taché in the legislative assembly proclaiming Québec's loyalty to the Crown: "Be satisfied we will never forget our allegiance till the last cannon which is shot on this continent in defence of Great Britain is fired by the hands of a French Canadian."[22] This attitude can be found in the philosophies of political leaders such as George-Étienne Cartier and Wilfrid Laurier who embraced the federal system as the only practical way to guarantee the survival of French Canada.[23] In the 1890s it led some French Canadians, including the historian Benjamin Sulte, to proclaim their imperial sentiments. Sulte, who had been a political secretary to Cartier, even translated "God Save the King" into French. But, even in its most imperial forms, this attitude still clung to its distinctive French heritage.

This strain of nationalism had its own military tradition to rival that of the Loyalists. It was encouraged by a succession of governor-generals who were sympathetic to the old world traditions of Québec and intent on enmeshing this heritage in mainstream Canadian culture. Thus, in 1827 Dalhousie sponsored the construction of a large monument in Québec City to honour both Wolfe and Montcalm.[24] French Canadians themselves were inclined to take a more exclusive view of their heroes. In 1860 the St Jean Baptiste Society erected a column designed by architect Charles Baillairgé to mark the centenary of the last French victory on North American soil and to inter the bones of soldiers subsequently unearthed at Lévis. But there was also interest in commemorating battles fought by Quebecers on the side of the English. The Québec Literary and Historical Society erected plaques to honour the defence of Québec against American troops during the Revolutionary War.[25] And there was widespread interest in commemorating the 1813 Battle of Châteauguay and its hero Lt Col Charles Michel de Salaberry.

The Battle of the Châteauguay has meaning on more than one historical level. There is what happened, and then there is the introduction of secondary layers of meaning as the event was interpreted in the context of later historical periods. In fact, the battle was characterized by a mixture of ineptitude, indecision, and courage often found in real life. In the fall of 1813 a large American force was poised to invade the Canadas. In the east an army of about three thousand men gathered on the border of Lower Canada under the command of General Wade Hampton.[26] About two thirds of this force, organized into two brigades then proceeded north with the objective of taking Montréal. As the Americans neared the Châteauguay River, Lieutenant Colonel de Salaberry gathered together a small force on the banks of the river to block the invasion route. Although a French Canadian by birth, de Salaberry had been serving as a regular officer with the British army since 1794 and had fought in the Napoleonic Wars. Returning to Canada in 1810, he had formed a French-Canadian militia unit called the Canadian Voltigeurs. The Voltigeurs, along with other militia and Indian allies – about three hundred men altogether – were under de Salaberry's command at the Châteauguay. The men dug in and awaited what they believed would be a major attack. Sometime before the arrival of Americans, however, Salaberry's men were reinforced by some fourteen hundred men led by Lt Col George Macdonell. Comprised in part of English-speaking militia units, the reinforcements did not man the line but instead took a supporting position behind de Salaberry's troops. Hampton, meanwhile, deployed one of his brigades against de Salaberry's position while sending the other under cover of darkness in a flanking manœuvre. The first brigade was stopped by the river and by de Salaberry's well-defended position, while the second brigade got lost in the dark and fog and failed to take the required objective. While this was happening, the Americans suddenly became aware of Macdonell's larger force and beat a tactical retreat. The actual engagement lasted about four hours and did not result in many casualties on either side.[27]

For the Canadians it was an important victory because the Americans did not again attempt to attack Montréal. For French Canadians in particular, however, it was significant in demonstrating that they were at least the equal of English Canadians in defending their homeland. Thus subsequent French-Canadian accounts tended to focus on the valiant stand of de Salaberry and his three hundred and overlook the support of Macdonell's men.[28] Sulte, in his *Histoire des Canadiens-Francais*, portrayed the battle as a modern Thermopylae, ranging three hundred brave men against seven thousand attackers. "L'antique bravoure de la race ne leur a pas fait défaut."[29]

Here again, however, public opinion about the significance of the site was not unanimous. In the case of Châteauguay, a revisionist view was advanced by a local anglophone historical society, the Châteauguay Literary and Historical Society, which contended that English Canadians under Macdonell had played a leading role in the battle. Some members even argued that de Salaberry had not been present in the action. Since it was the Châteauguay Literary and Historical Society which in 1888 petitioned Ottawa for a monument, its views had to be respected.[30] But it was Benjamin Sulte, then a senior official in the Department of Militia and Defence, who was put in charge of the monument and, considering the French-Canadian viewpoint, he found himself in a difficult position. Although he wished to prepare an inscription honouring the memory of de Salaberry and his men, he was instructed by the minister to compose a more innocuous text.[31] Hence the inscription that appeared on the monument declared feebly: "Here the invasion of Lower Canada was repulsed, and the enemy routed by the Militia of the Province."

Elsewhere in Québec the French-Canadian historical viewpoint was more successfully expressed. Perhaps its most prominent victory was at Fort Chambly where local enthusiasts managed to transform what was essentially a British fort into a shrine to the preconquest heritage. Fort Chambly is situated on the Richelieu River not far from Montréal on the site of a previous fort that had been built in 1665 to block the Iroquois invasion route from New York. The original fort was burned by accident in 1702 and rebuilt in stone. Following the capitulation of New France in 1760, the fort was taken over by a British garrison which greatly altered the stone structure during the American Revolution. After the garrison departed, in about 1850, the fort served as a military depot, but by the 1860s it was falling into decrepitude. In 1870, when it was transferred to the Department of Militia and Defence, the future of the fort, no longer strategically useful, was extremely doubtful.

By this time Fort Chambly had been discovered as an historic site, and a strong movement was building to develop it as a cultural resource. In 1865 Benjamin Sulte wrote a commemorative poem which appeared in Canadian and European journals, and the following year a Chambly journalist and antiquarian, Joseph-Octave Dion (1838–1916), publicized a plan to restore the fort in recognition of the bicentennial of the original wooden structure.[32] The fort attracted wide attention, and in 1875 the organ of the largely anglophone Canadian Antiquarian and Numismatic Society noted: "We believe that few could gaze at this time honoured ruin without feelings of emotion, and therefore deem it within the compass of our magazine to place on record a few notes especially as there has recently been shown some interest with a view of saving the ruins from oblivion."[33] The interest referred to was stimulated by Dion and reflected efforts to develop what was considered to be an important local resource.

But Dion did not wish merely to save the fort. He wanted to interpret its significance within a particular context of French-Canadian nationalist thought. Dion's nationalism was much narrower than the federalism of Sulte, Cartier, and Laurier. Instead of looking ahead to the salvation of French Canada within the federation of Canadian provinces, it looked back to an ideal society which it saw in preconquest Québec. Conservative in character, Dion was closely tied to ultramontane thought of the 1870s and 1880s and so he may be termed a clerical nationalist. His form of nationalism sought to promote traditional ideals of agriculture, language, and faith and was deeply wary of the modern trends being introduced by English-speaking Protestants. Like Ontario imperialism, then, clerical nationalism adopted a defensive stance toward the present and looked to the past for examples of heroic and patriotic resistance. Dion, as editor of the ultramontane *Nouveau-Monde,* was at the centre of this emerging nationalist ideology. Recently, a Canadian Parks Service historian described his belief in the following terms: "Il s'accroche au modèle de la chrétienté rurale en terre laurentienne. Influencé par les courants de pensée ultramontain et agriculturiste d'Europe, il est un des défenseurs de cette nouvelle idéologie. Ce nationalisme des ultramontains, dont Dion fait partie, est réduit à sa dimension culturelle; défense de la religion, de la langue et des institutions."[34] In this context, forts suited conservative Québec nationalism just as they suited its Ontario counterpart. Here, though, Dion was looking at the past selectively, ignoring the British history of the fort while focusing on its preconquest significance.

Ironically Dion acquired some of his initial ideas for the fort from contemporary France where he had been involved in preservation causes. And in 1873 he launched a campaign there in an attempt to raise funds for the restoration of Fort Chambly.[35] But this effort produced no results. He had greater success with an 1879 campaign in Québec to raise funds for a statue to de Salaberry. He was then able to capitalize on the enthusiasm surrounding the investment of this monument and he inaugurated another drive for the restoration of the fort. This time he approached the Canadian government through his local member of Parliament.[36] Ottawa responded favourably to his proposal, making him honorary curator and directing the Department of Public Works to prepare preliminary estimates for the stabilization of the structure. In its report presented in November 1881, the Chief Architect's Branch stipulated that one thousand dollars was necessary to carry out vital repairs.[37]

The estimates received the necessary approval, and the repairs were carried out under the supervision of Dion, who strove to keep the work compatible with the original design. The work could not be completed within the allotted budget, however, and in 1883, through Dion's urging, Parliament approved the spending of an additional one thousand dollars. Dion then increased his minimum requirements, and by 1884 almost five thousand dollars had been spent on restoring the walls and building new structures. Among the latter was a small residence in the parade ground which Dion later developed into a museum to display curios, not only from the fort's history, but also reminiscent of traditional Québec life. A rebuilt archway to the fort was inscribed with the names of battles and wars associated with the French and British forts on the site. In 1887 the property was transferred to the Department of the Interior and placed under the care of the Department of Public Works with help from the Department of Militia and Defence. That year a memorandum from the Minister of the Interior stated that the "old fort ... [had] been placed in a fair state of repair, with a view to its preservation as a very interesting memorial of early Canadian History."[38] Dion, it seems, had assumed total control for the interpretation of the fort's significance. He and his successor, L.J.N. Blanchot, who replaced him in 1916, looked upon their charge as a sacred trust from the French-Canadian past. They encouraged pilgrimages by religious and nationalist organizations who came to regard the fort as a shrine to traditional Québec culture.[39]

There were also large heritage developments that were inspired by other concerns besides nationalism. Fort Lennox, located on the 210-acre Île-aux-Noix a few miles up the river from Chambly, was a natural tourist attraction whose historic potential was developed for commercial rather than ideological ends. It was used by French troops during the seventeenth-century French-English colonial wars, and by British troops during the American Revolution and the War of 1812. The British army established Fort Lennox here in the 1820s when a group of fine ashlar stone buildings, surrounded by an earthwork wall and moat was constructed. Obsolete as a defensive structure almost as soon as it was built and inconvenient to road transportation, its function as a military depot was gradually assumed by the establishment at St Jean, which was closer to Montréal, and by the 1860s it was virtually abandoned. But useless as it was to the military, the fort and meadows of Île-aux-Noix formed a natural attraction for tourists who came by boat in ever-increasing numbers from the United States and Montréal. While aware of this new function, the Department of Militia and Defence, which had inherited the island from the British, seemed reluctant to dispose of it permanently and instead leased it to private entrepreneurs who effectively turned it into an historic park.

The Richelieu River Navigation Company obtained the first lease in 1899 but failed to exploit the potential of the island, and two years later the lease passed to B.V. Naylor who ran a cruise boat along the river between Lake Champlain and St Jean. Naylor developed the island as a stopover for excursionists.[40] He persuaded the military to allocate funds for the repair of the fort's buildings and through the first decade of this century he carried out some major renovations: he built a bridge over the moat, repaired walls, and added a new roof to one of the buildings. To enhance the attraction of the old fort he installed a museum in the former men's barracks, which he gradually expanded over the years. A large part of the museum collection consisted of small articles found on the island, such as regimental buttons, coins, and arrowheads, but it also included other artifacts of a military nature, some of which were obviously American in origin. The exhibit included a sword taken at the 1813 Battle of Chesapeake, Boston, a cavalry carbine allegedly used at the Battle of Eccles's Hill in 1860, as well as an assortment of cutlasses, rifles, and a flintlock. Two prized items were a seventeenth-century coin (presumably of European provenance) and a miniature Bible inscribed to George Washington.[41]

In the Maritimes at the turn of the century nationalist perspectives were not nearly as well focused as in Ontario and Québec. While traces of imperialism and even French-Canadian nationalism can be detected in Maritime historiography of this period, there did not appear to be a well-defined sense of historical purpose. Nevertheless, there was considerable antiquarian interest in the past for its own sake. These tendencies shaped a rather loose interpretation of historic sites in the region. Rarely were sites as didactically presented as Lundy's Lane. And the preservation of heritage properties sometimes had rather confused objectives. This lack of a dominant local nationalist perspective helps to explain developments at Annapolis Royal in Nova Scotia, where there is a cluster of historic sites, one of which is the site of Champlain's original 1604 habitation. After the restoration of Acadia to France in 1632, a settlement was established at the site of present day Annapolis Royal and named Port Royal. A wooden fort was built there which finally fell to the British in 1710. At this time the settlement and fort were renamed after Queen Anne of England. New fortifications were constructed, and in 1797 a brick officers' quarters was built. Annapolis was thus important from three perspectives: as the site of one of the earliest European settlements in North America (Habitation Port Royal), as the former *chef lieu* of French Acadia (Port Royal), and as a British garrison and collection point for American Loyalists fleeing the revolution (Fort Anne). Before the 1920s it was usual for these three closely situated sites to be mixed together, and Fort Anne was popularly regarded as the birthplace of civilization in North America. The relevance to subsequent Canadian development was ignored.

The fort became the collective symbol of the various historic stages of the place because it alone survived from the past. The fort, then, became the focal point for the local historical community. The British had virtually abandoned the fort in the 1830s, but it was not transferred to the Canadian government until 1870 when, as ordnance property, it was placed under control of the Department of Militia and Defence. Strategically unimportant and from the beginning virtually useless to the military, the department nevertheless made some efforts to preserve the old landmark. In 1899 the property was leased to a local management committee, formed under the name of the Garrison Commission, which received regular allocations for the maintenance and supervision of the property. The Garrison Commission was not an agent of the government but assumed a separate identity and took important initiatives. It advised the government to lease part of the land to farmers and suggested ways in which the fort could be restored. Funds were obtained from the federal government, a municipal appropriation, and "the occasional entertainment."[42] In 1902 the commission requested F.B. Wade, the local member of Parliament, to approach the government for an increased appropriation.

The Garrison Commission formed, in effect, a nascent local historial society intent on promoting the rich historical associations of the area through the physical resource of the fort. In 1904, in conjunction with the Nova Scotia Historical Society, the commissioners launched a fund-raising drive for a monument commemorating the founding of Port Royal. A stone memorial was subsequently erected on the grounds of Fort Anne and inscribed with the following legend:

To the illustrious memory of Lieutenant-General Timothe Pierre du Gaust, Sieur de Monts, the pioneer of civilization of North America, who discovered and explored the adjacent river, AD 1604, and found on its banks the first settlement of Europeans north of the Gulf of Mexico.

The Government of Canada reverently dedicates this monument, within sight of that settlement, AD 1904.

The federal government was encouraged by the local organization to give a kind of official recognition to the area's history. Parliament approved spending three thousand dollars on this project, and a further two thousand dollars was appropriated in 1904.[43] Subsequently, however, the Department of Militia and Defence attempted to divest itself of responsibility for the fort. While the Nova Scotian, Frederick Borden, was minister of Defence, the department seemed reasonably sympathetic to the problems of picturesque old forts in the Maritimes. But after his replacement by Sam Hughes in 1911, the military assumed a more hard-nosed stance. In August 1913 the military secretary pencilled a minute on a department memorandum on the question of funds for repairs to Fort Anne. "No! the Minister does not approve of expending Militia Votes on old abandoned forifications – if these are to be maintained for sentimental or historic reasons, it must be done through special appropriations for the purposes."[44] And, when the Garrison Commission tried again in 1914 to have the local member of Parliament lobby the department for funds, the military secretary indicated that this source of support was no longer available.[45]

Cut adrift in this way, the Garrison Commission cast around for a new sponsor, and in February 1916 a joint meeting of the Town Council, Board of Trade, and the commission was held to consider "the best means of having the old historic garrison property in this town taken care of."[46] The meeting was principally concerned with considering a proposal by L.M. Fortier, a retired official of the Department of the Interior and a leading member of the Garrison Commission, to have Fort Anne taken over by the Dominion Parks Branch as a national park. Fortier argued that this was the only expedient option left as the government already owned the property, and national park status would provide a regular appropriation. The development of the fort by strictly local means was out of the question. Fortier's proposal was endorsed by the meeting and, with the co-operation of parks officials in Ottawa, Fort Anne National Park formally came into being. Fortier was named honorary curator, and the Garrison Commission, reincarnated as the Annapolis Royal Historical Society with Fortier as president, assumed paternal responsibility for the fort. There was some criticism about having the site taken over as a national park and renaming it Fort Anne. But generally there was satisfaction over the plan to develop Fort Anne as a park largely because it would provide a landscape evocative of a range of historic associations and a recreational area for local citizens and tourists.

Undefined historical associations also governed the development of Fort Beauséjour in New Brunswick. Begun as a French fort in Acadia about 1750, it was later taken over by the British who rebuilt and renamed it Fort Cumberland. Earthworks and a brick powder magazine were the principal remains after the British left, and in 1875 the Department of Militia and Defence transferred the property to the Ordnance and Admiralty Lands Branch of the Department of the Interior.[47] Although in New Brunswick, the site was of interest to nearby Amherst, Nova Scotia, and its preservation was adopted as a cause by the Nova Scotia preservation movement. Both English and French associations were considered to be equally important, however, and local heritage entrepreneurs, eager to develop the site as a tourist attraction, were content to refer to the British-built fortifications as Fort Beauséjour. As it was government property, Ottawa was persuaded to assume responsibility for its preservation as an historic site. In 1902 the Ordnance Lands Branch fenced off the fort. The branch proposed restoring some of the old structure but as special funds were not available the project was abandoned. Then, in 1913, a military engineer inspected the site on behalf of the branch and drew up a plan for a new fence and the restoration of the powder magazine.[48] Parliament approved spending five thousand dollars on the property, but the outbreak of war caused the project to be shelved.

A more ambitious attempt to develop an historic property in the Maritimes at the turn of the century took place at Louisbourg, Cape Breton. Although the historic ruins were not on public property, the sheer scale of the problem led to government involvement. Typically, the significance of the site was interpreted very broadly. The Fortress of Louisbourg had been important for defending the Atlantic approach to New France and its collapse in 1758 was an important prelude to the fall of Québec the following year. Although systematically destroyed by British forces and subsequently abandoned as a settlement, enough of its casemates and ruined walls remained to suggest its former grandeur. Francis Parkman described its romantic setting in his epic history *Wolfe and Montcalm*. "Here stood Louisbourg; and not all the efforts of its conquerors, nor all the havoc of succeeding times, have availed to efface it. Men in hundreds toiled for months with lever, spade, and gunpowder in the work of destruction, and for more than a century it has served as a stone quarry; but the remains of its vast defences still tell their tale of human valor and human woe."[49] It was this former glory, not the ignominious defeat, which subsequent preservationists were eager to portray. Hence, although the historical significance of the site was justified in terms of its importance to British imperial history, it was the existence of the fortress prior to its capture that local entrepreneurs wished to recreate.

In 1903 a retired Indian navy captain, D.J. Kennelly, purchased the property on which the ancient bomb proofs and casemates stood – the only extant part of the fortress – and undertook to develop the ruins as a privately run historic attraction. He propped up the sagging brickwork with timber and patched the mortar where it was crumbling. Apparently he built a restaurant within the ruins and charged twenty-five cents admission to the site.[50] In order to finance further plans for development, in 1904 he organized the Louisbourg Memorial Association to receive charitable donations. He canvassed the messes of British regiments present at the second siege of the fortress and even received a five thousand dollars grant from the Canadian government. One witness has described him as an eccentric old man,[51] but he obviously possessed some sense and influence because in 1906 he convinced the government of Nova Scotia pass an Act to incorporate the trustees of the fortress and old burying ground as an "Historic Monument of the Dominion of Canada" and as a public work. As further testimony to his organization's credibility, the trustees included Baron Strathcona and Mount Royal, Earl Amherst, Viscount Falmouth, Everett Pepperrell, and Sir Frederick Borden.

The major obstacle to the development of the old fortress as a national shrine was access. The town of Louisbourg was at the time remote from the rest of Nova Scotia, and the site of the old fortress was even more inaccessible. The first project of the Louisbourg Memorial Association was, therefore, to build a road from the town to the historic property. Previous to this, Kennelly had involved himself in an elaborate and futile scheme to connect Louisbourg with the mainland of Canada by a chain of railways and ferries between Louisbourg and Ontario called the Canadian Atlantic Ferry. The only part of this scheme to be implemented, however, was the acquisition of a railway right of way through part of the fortress site, and when Kennelly died in 1912 he left behind a hopeless legal tangle concerning the ownership of the Louisbourg property.

It is clear from the way Kennelly promoted the site by enlisting support from British regiments and the descendant of the eighteenth-century governor of Massachusetts that he interpreted its significance in terms of imperial history. There was at the turn of the century, however, a mounting body of public opinion concerned with promoting the French-Canadian aspect of the site. Men like John Bourinot and Pascal Poirier, who had visited Louisbourg in 1902, saw the fortress as a symbol of the former greatness of French-Canadian civilization in North America.[52] Bourinot had reported the significance of Louisbourg in this context to the Royal Society as early as 1891 in his paper, "Cape Breton and its Memorials of the French regime."[53] The interpretation of the French significance of the site assumed greater local importance with Kennelly's successor, J.S. McLennan.

In the 1880s, McLennan (1853–1939) came to Sydney, Nova Scotia, from Montréal to work for a coal company and in 1899 presided over the formation of the Dominion Iron and Steel Company, becoming its general manager. Later he resigned from this salaried position and bought a local newspaper. He gradually slipped into semiretirement and became engrossed in the history of the ancient fortress at Louisbourg. A wealthy and cultivated man – he was a graduate of McGill and Cambridge universities – he was well suited to this task. He visited principal archives in Canada, the United States, France, and Britain and amassed a large number of documents and artifacts relating to Louisbourg. McLennan was interested not in the downfall of the fort but in its life, and his *magnum opus, Louisbourg, From its Foundation to its Fall, 1713–1750,* completed in 1914 but not published until after the war, celebrated its pre-British glory. In researching its early history, McLennan had uncovered original plans for the fortifications of Louisbourg and as early as 1908 he argued that it would be possible to reconstruct the old fortress. In an address that year to the Nova Scotia Historical Society, "Louisbourg as a National Charge," he proposed that the site of the ruins be acquired by the Canadian government and restored as a national monument.[54] McLennan was undoubtedly influenced by plans, begun that year, for the reconstruction of Fort Ticonderoga in New York State by Stephen Pell, although he realized that Louisbourg was on such a scale that few individuals could afford to sponsor its restoration. Nevertheless, in promoting the development of the site as a national monument, it was its significance to French rather than the subsequent British period, that he chose to emphasize.

In the Maritimes the heritage movement was not so well organized as in Ontario and Québec but it was still an emerging force. The Nova Scotia Historical Society had been active since the 1890s, mainly centred in Halifax but managing to promote causes as far afield as Louisbourg. The Nova Scotia Archives was also generally concerned with heritage issues, as was the curator of the provincial museum. In New Brunswick the heritage movement was concentrated in loyalist groups which formed to promote the centenary of the province. In the west there were a few fledging historical societies, most notably the Manitoba Historical and Scientific Society. More often, though, it was the provincial librarian who was concerned with provincial historical matters.

The heritage movement assumed a national outlook through the co-ordinating efforts of the Royal Society of Canada. The society had functioned since the 1880s as a national cultural agency promoting scholarship in both the arts and sciences. It was organized into three sections: the first represented French language letters, the second was concerned with literature and history in English, and the third was devoted to science. Each section elected its own members which were limited by a quota. They brought together at the annual meetings and through the published *Proceedings and*

Transactions the representative work of Canadian scholars. In the late nineteenth and early twentieth centuries the membership was not dominated by academics as it is today. Before the 1930s academics comprised only a small fraction of the total membership, especially in the first two sections. But the society brought leading amateur scholars together in a forum devoted to promoting higher learning in the country.

As its membership included many prominent members from local historical societies, as well as many local historical societies as corresponding members, the Royal Society came to devote more and more attention to heritage issues. The 1891 *Proceedings and Transactions* contained a report by J.G. Bourinot on "Cape Breton and its Memorials of the French Regime" and a plea for the preservation of the Château de Ramezay in Montréal. That year the council of the Royal Society, on the prompting of the Numismatic and Antiquarian Society, proposed that the Château de Ramezay be made "a national repository of those memorials and archives, which will illustrate the history of Montreal generally."[55] The Royal Society pointed to efforts being made to preserve nationally significant buildings in the United States as further reason for Canada to recognize its own nationally significant heritage.

In 1901 the Royal Society was contacted by the British organization "The National Trust for Places of Historic Interest or Natural Beauty" which was soliciting aid for the development of antique places in the country. The goal of the National Trust, which at that time had existed for five years, was to acquire property from individuals and then administer this real estate for the benefit of the British public. A quasi-public institution, it lacked a regular income from the government and relied on an endowment and public subscriptions. In an effort to broaden this financial base it decided to ask expatriate Britons overseas for donations and endeavoured to form associate societies in the United States, Canada, and elsewhere in the empire. This campaign, led in North America by C.R. Ashbee, met with some success, but it also stimulated preservation activities in Canada by inspiring the Royal Society to attempt to emulate what the National Trust was doing in Britain. In reply to Ashbee's communication asking for support, the council of the society passed the following resolution: "The council believes that the best way in which the Royal Society can assist the very desirable object aimed at by the National Trust is to form a Canadian Committee for the preservation of places, of scenic and historic interest within the Canadian Dominion as a part of the British Empire."[56] The Committee for the Preservation of Scenic and Historic Places in Canada was formed in 1901 as a consequence of the resolution.

From the beginning the committee was concerned with the preservation of historic rather than natural landscapes. This is evident from the background of its members who included Sir James LeMoine, Benjamin Sulte, Mr Deville, Hon. Pascal Poirier, John S. Willison, Lt Col George T. Denison, and John Bourinot. These members already had between them considerable experience in the heritage movement. LeMoine was president of the Literary and Historical Society of Québec and had written extensively on landmarks in that city. Sulte was a noted Québec historian active in preservation issues in the province but especially in the Trois Rivières area. Sen. Pascal Poirier from New Brunswick was interested in historic Acadia, particularly Louisbourg, and Denison, as we have seen, was concerned with the history of the Loyalists and had been involved in the commemoration of the battlefields at Crysler's Farm and Lundy's Lane.

The committee was able to use the prestige of its members both in the national cultural community and among local heritage groups to bring hitherto local concerns into a national sphere. It therefore became a driving force in the heritage movement, inspiring new efforts among local groups and raising the expectations of well-established societies. Further, as an Ottawa-based society which included senior government officials as well as the cream of the intellectual élite, it was well positioned to influence greater government involvement. The committee soon began to live up to its potential. In 1902 it sent Poirier to Louisbourg to report on the condition of the ruined French fortress. This led to Poirier's involvement in the movement to restore Louisbourg, and to the committee's consideration of the problem of preserving historic buildings. Following Poirier's address to the Royal Society that year, that body passed a motion introduced by the senator: "That the Royal Society would respectfully memorialize the Dominion Government to take steps for the acquisition and preservation of historic sites, buildings and places of interest which have a national importance."[57] The following year the committee asked the Royal Society to pass a resolution drawing the attention of various governments in Canada "to the importance of preserving historical monuments, sites, buildings, archives and relics throughout Canada in view of the constant and increasing danger of their disappearance."[58]

By 1907 the activities of this committee and local historical societies in promoting historic sites had risen to a level sufficient to warrant the formation of a separate national body. Although this widespread public interest was the culmination of a fairly long process, enthusiasm was further sparked that year by the publicity surrounding the plans to develop the Plains of Abraham. This event aroused considerable interest across the country for it pointed to the potential of historic landmarks as scenic and cultural attractions. The time was thus ripe for the creation of a national heritage body, and it was with this objective that in May 1907 the Royal Society sponsored a convention in Ottawa attended by members of the society and various historical organizations. Among those present were: Colonel Wood, past president of the Québec Literary and Historical Society, who chaired the meeting and was subsequently elected president of the new society; Col Ernest Cruikshank of the Ontario Historical Society, who was elected vice-president; as well as Dr Robert Bell; Rev. F.G. Scott; Benjamin Sulte; W.W. Campbell; James H. Coyne; George Drummond; W.D. Le Sueur; W.D. Lighthall; and Mrs J.H. Thompson.

Called the Historic Landmarks Association, the new society had as its first priority the development of the Plains of Abraham as a national historic site. But following the establishment of the Québec (subsequently renamed the National) Battlefields Commission early in 1908, it changed focus and set out to be a clearinghouse for information about historic sites across the country. At this point it had little, if any, capacity to implement projects itself. Its strength remained in the local societies which were willing to participate in a master plan which it devised and in its ability to lobby government agencies for money. For this reason the Historic Landmarks Association declined to become involved in costly preservation projects and instead set out to recognize significant sites across the country by means of a national inventory. Its procedure was to encourage local societies to nominate the most significant sites in their jurisdictions and then publish them in an annual report. Its criteria for the designation of historic sites were deliberately loose, allowing grass roots organizations considerable latitude in their selection. "**What is a landmark?** A landmark is anything preservable which is essentially connected with the great thoughts and deeds that once stirred our life and still stir our memory. It may be a monument set up by pious hands; a building, a ruin, or a site; a battlefield or fort; a rostrum or a poet's walk; any natural object; any handiwork of man; or even the mere local habitation of a legend or a man."[59] Such a site would assume national significance merely through the sponsorship of the Historic Landmarks Association. It did not have to be significant to a national history. Nationalist ends, then, were being deliberately downplayed.

Given the importance of nationalist sentiment in promoting sites such as Lundy's Lane, and the prominence of figures such as George Denison in the Royal Society, it is remarkable that the association did not justify historic landmarks as symbols of a unifying Canadian heritage and instead placed so much emphasis on local significance. The explanation for this behaviour was not difficult to find in 1908, however, for by this time there were numerous conflicting nationalist views revealed in the process of selecting and interpreting historic sites, as many local groups promoted historic sites on the basis of their significance to a particular national history. It is ironic, then, that as a national body the Historic Landmarks Association had to evade nationalist ideology in order to avoid conflict. Competing nationalisms, while central to the development of a national heritage movement, presented problems for a national program commemorating local historic sites.

The capacity of the association for action was so limited that it could not even publish an annual report until 1915. But with the introduction of the annual report in 1915 it became clear that the Historic Landmarks Association had greater ambitions than merely collecting information and aspired to become an agency through which official recognition could be bestowed on worthy sites. To effect such a program it needed the co-operation of the federal government to buy and put up the commemorative markers which it envisaged. "This Association desires to gather from all parts of the Dominion of Canada, all the knowledge available regarding each site or case it is proposed to mark – obtain verification of the same from documents in the Dominion Archives and other reliable sources, submit reports from each province to the Council, which will then consider the merits of each application and, when desired, recommend to the Government for approval."[60]

In effect, the Historic Landmarks Association saw itself acting as an advisory board to the government, screening applications from local organizations for commemorative markers. It is likely, too, based on subsequent events, that the association envisaged the enactment of some kind of heritage legislation to protect these sites from destruction. For a while it appeared as if the government would collaborate with this objective: the commissioner of the National Parks Branch was a member of the association, an inventory of historic sites being undertaken by his office shared information with the association, and in 1920 the branch subsidized the publication of the Historic Landmarks Association report. But the Historic Landmarks Association was a dead end. When in 1919 the government appointed its own body to advise on historic sites, much of the *raison d'être* of the association was lost, and in 1922, following the election of Lawrence Burpee as president, the association transformed itself into the Canadian Historical Association and adopted a new set of objectives.

THE CONSERVATION MOVEMENT AND THE NATIONAL PARKS SERVICE, 1911–19

In 1919 James Bernard Harkin (1875–1955), the commissioner of national parks, was one of Ottawa's more promising civil servants. A former newspaperman and political secretary to Clifford Sifton while he was minister of the Interior, Harkin had been handed the task of forming a national parks service in 1911. He undertook this with characteristic energy and by the outbreak of the war had made the national parks service into an important branch of the Department of the Interior. To understand how historic sites became a part of Harkin's mandate we need first to know something of the nature of his program and the larger conservation concerns which supported it.

Ottawa's commitment to national parks grew out of earlier conservation initiatives. The national forest reserves were viewed as having other assets than the commercial value of their trees. There was a wide body of opinion, well represented in the United States by John Muir, but well articulated in Canada by nature writers such as Ernest Thompson Seton and Charles G.D. Roberts, that the wilderness forest should be protected for recreational areas. At the turn of the century urban dwellers demonstrated an unparallelled enthusiasm for nature, founding outdoor clubs and bird-watching societies. Moreover, the urban middle class popularized such outdoor pursuits as hiking, camping, and canoeing, activities that an earlier generation would have considered mere hard work. While idealists such as John Muir and his acolytes in the American Sierra Club promoted wilderness parks for their moral and scientific value, even philistines in the government could be convinced of the commercial potential of parks as tourist attractions.

In Canada, wilderness reserves were almost entirely concentrated in the Rockies, inspired by both the practical and aesthetic streams of the conservation movement. The completion of the transcontinental railway made many aware that wilderness areas were in imminent danger of being alienated from public use. At the same time there was growing awareness of the tourist potential of natural scenery. The CPR was therefore only too happy to have the Department of the Interior assume control of a few acres of land surrounding a hot spring near present-day Banff.[61] Canada's first national park, Rocky Mountain Park, established by ordinance in 1885 and by the Dominion Act in 1887, became a mecca for outdoor enthusiasts and a boon to the railway which transported them there. By 1911 a number of other park reserves had been established in the Rockies along with large areas of forest administered by the forestry branch. That year the minister of the Interior, Frank Oliver, set out to provide for the better administration of park and forest reserves by proposing a new act. Legislation identified large areas of Crown land as forest reserve "for the maintenance, protection and reproduction of the timber ... for the conservation of the minerals and the protection of the animals, birds and fish therein, and for the maintenance of conditions favourable to a continuous water supply." The act also authorized the establishment of Dominion Parks, stipulating that they were to "be maintained and ... made use of as public parks and pleasure grounds, for the benefit, advantage and enjoyment of the people of Canada."[62] The existing parks were considerably reduced in size. Surplus land was transferred to adjacent forest reserves, but provision was made for future park expansion from the present reserves.

The Dominion Parks Branch was established following the passage of this act to look after the development of national parks in tandem with the forestry branch which would confine itself to the protection of forest reserves. James Harkin, then a senior official in the department, was invited by Oliver, the outgoing Liberal minister, to head the new branch. At this point the possibilities for park development were severely restricted by the act. Moreover, there were conflicting pressures regarding the direction park development was to take. On the one hand, there were those, including the minister, who believed that parks were a natural resource, capable of being exploited like a forest. On the other hand, there were those, including members of the influential Alpine Club of Canada, who preferred to see parks kept in their pristine state. In the ensuing years Harkin managed to implement a policy that would largely satisfy both groups. His initial difficulties lay in the huge scope of his task, and the limited resources which he had at his disposal. Years later he wrote in a memoir:

> My first problem on taking office was the economic one. How was I to get the money for developments that were immediately necessary? To restore the disappearing wild life, an efficient game protective service with a code resembling that of our Mounted Police should be built up ... Hundreds of miles of new trails and forest telephone lines were needed at once ... Motor roads within the parks could not long be denied.
>
> All this would cost money – a great deal of money. How could the hard-headed member of the House of Commons be persuaded to increase park's appropriation?[63]

The crux of his problem was that he needed more money in order to make parks more popular and yet to obtain this money he needed to make parks popular.

He approached the problem by launching a campaign to sell the idea of parks. In press releases, memoranda, and annual reports of the department, Harkin trumpeted the theme that parks were good for the country. Prior to 1914 the principal strands of his argument were that parks were cost effective because they attracted tourist dollars and were of benefit to the physical and moral health of the nation. He developed the first line of his argument by amassing data demonstrating the economic potential of tourism. The CPR collaborated in this endeavour by estimating "the amount of money attracted to Canada annually by the fame of the Rockies at about $50,000,000."[64] Harkin stressed the importance of the national parks to this industry, summing up his argument with the quip "nothing attracts tourists like national parks."[65] Harkin then cited dozens of scholarly studies to show the importance of parks for society's physical and mental well being. "The most important service which the parks render," he wrote, "is in the matter of helping to make the Canadian people physically fit, mentally efficient, and morally elevated."[66] Embedded in Harkin's sales pitch was the belief that it was the duty of the federal goverment to develop parks for the economic and social benefit of the nation.

Harkin was enormously successful as a salesman of the national park idea. This helped him personally for he was able to convince the newly elected Conservative Borden government that he was a national expert on the subject rather than just another Liberal placeman. He developed an excellent rapport with the most influential minister of the Interior in the Borden administration, Arthur Meighen, who held the portfolio between 1917 and 1920. While the parks branch budget was trimmed to allow only minimal operating expenses during the war, the branch received steadily increasing appropriations after 1918. Further, Harkin managed to improve the extremely weak 1911 legislation which had tied parks to forest reserves. In 1913 an amendment was passed to the 1911 act which no longer made it a requirement for a park to be created from a forest reserve.[67] Consequently, the branch managed to increase the area of its Rocky Mountain parks and looked to establishing new parks to the east.

Despite the success of his campaign to sell parks as a national resource, Harkin was vulnerable in being unable to reach the bulk of the Canadian public. The national parks were concentrated in the Rockies, beyond the reach of all but the more affluent Canadian tourist. The parks branch needed a national system of parks in order to establish a national presence and offer the promised recreational facilities to all Canadians. Unfortunately, there were a number of major obstacles to establishing such a system. The original national parks had been formed out of undeveloped federal Crown land. Such property, offering the requisite scenery, was difficult to acquire in the more populous east. Moreover, the whole orientation of both his department and his branch was firmly rooted in the west. Much of Harkin's budget was, therefore, committed to developing the obvious money-making parks in the Rockies, and there was little left over for costly acquisitions in the east.

During the early years of his administration, Harkin groped for expedient solutions to this problem of creating a greater national presence. In his annual report for 1913, under the heading "New Lines of Development," he suggested making an official from his office available to give expert advice to cities willing to set up playgrounds.[68] In this way, he argued, outdoor activities could be better co-ordinated along national standards. Another approach was to suggest that municipalities be encouraged to preserve open spaces from surrounding land for camping and other outdoor pursuits. Ideally, Harkin would have liked to have been given the mandate for developing these suburban parks, but he was unsuccessful in having either of these too obviously municipal or provincial responsibilities handed over to him.

The commissioner was more successful with yet a third proposal for new development presented in 1913. In relation to the campsite proposal, Harkin suggested that some parks might be created at points of historic interest. Perhaps features from human history could be preserved as focal points and attractions to the suburban park. Like the natural parks, he rationalized the creation of historic parks by his doctrine of usefulness. "It would be doubly beneficial if these historic spots were not only properly restored and marked but they should be used as places of resort by Canadian children who, while gaining the benefit of outdoor recreation, would at the same time have opportunities of absorbing historical knowledge under 'conditions that could not fail to make them better Canadians'"[69] This proposal was accompanied by an initiative of the branch in 1914 to make "a general survey of the historic sites of the Dominion with a view to preserving them as national landmarks and monuments."[70] That same year the branch acquired Fort Howe in Saint John, New Brunswick as a national park and three years later, Fort Anne. In this way the idea of the historic park was born.

Initially, these historic resources were viewed from the particular perspective of the parks service. They tied together notions of leisure and scenery and those of education and history. Apparently only marginally interested in the historic ruins for their own sake, Harkin seemed more interested in them as a rationale for creating a surrounding recreational area. Fort Howe was not a very important historic site, yet it was used to justify an acquisition of some otherwise undistinguished property for a park.[71] Fort Anne had better credentials as a historic site, but in Annapolis Royal local opposition to having the site developed as a park pointed to potentially contradictory objectives. In a letter published in the local newspaper, the well-known local historian Judge A.W. Savary explained: "AND FORT ANNE NATIONAL PARK! What a hybrid combination! A fort with its bastions, ravelins, glacis, moat, and a park with its shade trees, avenues and drives; the two are absolute [*sic*] incompatible."[72] Nevertheless, the belief that historic and natural landscapes could be treated together was widely held at this time. Preservation groups in England and the United States regularly combined heritage with scenic conservation,[73] and the landscaping of the Plains of Abraham had received widespread approval in Canada.

The association of historic sites with landscape preservation provided the national parks branch with a relatively easy way to extend the parks system across the country. As early as 1913, therefore, the branch was sending the following request to knowledgeable groups and individuals. "The Commissioner of Dominion Parks is at present very much interested in this matter of having historical monuments, etc., in the Dominion preserved and if you could, without too much trouble to yourself send to him a short resume of the facts in this connection ... which will enumerate places historically worthy of preservation, I think it might result in their preservation in the future."[74] The war put an end to this scheme but by 1919 the branch was taking steps to preserve the sites in Saskatchewan of Fort Pelly and the burial ground of the men who fell at the Battle of Fish Creek, "with the intention of sometime establishing them as historical parks or monuments."[75] And so, by the end of the war the branch was in the historic sites business, but only tenuously as it had little money with which to develop these newly acquired resources.

The ambitions of the branch to expand the parks system through the acquisition of historic sites coincided with the department's growing concern with what to do with some of the heritage properties under its control. In 1919, William J. Roche, the minister of the Interior was particularly concerned with the fate of the old western fur trade posts. Previously in 1914 a convention of Ontario historical and patriotic societies passed a resolution to petition the minister of the Interior to develop a forty-acre site near Thorold as a national battlefield park commemorating the Battle of Beaver Dams.[76] These issues led the minister in 1919 to ask Harkin to formulate a proposal for a heritage policy to be administered by the department.

By this time Harkin and his staff had acquired enough experience with heritage properties to know that they required different treatment than natural parks. While the branch possessed the operational capacity to physically care for sites, it lacked historical expertise to deal with the special problems posed by heritage resources. First was the problem of criteria. There were so many well-publicized historic sites in the Maritimes, Québec, and Ontario that even with an ample budget the branch could not afford to acquire them all. The weakness of Fort Howe as a national historic site must have helped convince Harkin of the need to set priorities to avoid succumbing to short-term expediency. Yet with so many local groups lobbying for their own heritage projects, who could select the most important sites for preservation? Second was the need for reliable historical advice concerning the commemoration and development of historic sites. Harkin took Judge Savary's barbs about Fort Anne National Park seriously, and the branch came to believe that it had been misled by the local historical society in naming the park after the British and not the French establishment.

In a memorandum to the minister written in March 1919, Harkin proposed a scheme for establishing national historic sites across the country that would resolve some of the earlier problems his branch had encountered. "To overcome the difficulty of determining which sites are truly of Dominion-wide concern, I would suggest that an honorary board or committee, following the line of the Wild Life Board, be appointed, composed of men from all parts of the country who are authorities on Canadian history, to advise the Department in the matter of preserving those sites which pre-eminently possess Dominion-wide interest."[77] This document provided the basis for a formal government heritage program. Henceforth, responsibility for the identification and preservation of national historic sites would rest with the parks branch, which would be advised by the Historic Sites and Monuments Board of Canada. As noted by Harkin, there were precedents for this arrangement. Besides the Wildlife Board, which existed to inform the minister of the Interior on the regulation of fish and game, there was also the Historic Manuscripts Commission which advised the dominion archivist on the preservation of documents. The existence of these advisory bodies reflected the new faith in the role of the expert to provide objective and rational advice for the enactment of government policy. Cynics might also point out that these boards could be presented to the public as autonomous agencies whereas in reality they were merely advisers without authority whose recommendations could easily be disregarded. But in 1919 the nature of the board's role had scarcely been defined and its true function would only emerge in the ensuing years.

While the national parks service's new responsibilities and the creation of the Historic Sites and Monuments Board resulted from branch ambitions and department concerns, the establishment of this new program by the government was the result of a long series of events involving a number of different groups. Local heritage groups, the agitation of local historical associations, and the emergence of a national heritage movement were all reasons for the government becoming involved in the preservation of historic sites. The expectations of these groups combined with those of the parks branch to help define the conflicting nature of the historic sites program in the ensuing decades. Only clear direction from Ottawa in the form of a policy could resolve some of these contradictions, but such direction would be a long time coming.

CHAPTER TWO

Birth of a Program, 1919–23

The implementation of a federal program of heritage conservation was in 1919, as it is today, an enormous undertaking, yet James Harkin had few means at his disposal to inaugurate this new activity. His main resource in the first phase of the program's history was an advisory committee, the Historic Sites and Monuments Board. It consisted of leading figures of the heritage movement: E.A. Cruikshank and Benjamin Sulte from Ottawa; James Coyne from St Thomas, Ontario; W.C. Milner from Halifax; and W.O. Raymond from Saint John, New Brunswick. The effectiveness of this group was limited, however, first by internal discord and then by its isolation from the government bureaucracy. By 1920 Harkin had designated staff to direct the program, but these measures were too meagre to have much effect.

Fundamentally, the government was unwilling to commit funds or legislation to the new heritage activity. Without such basic underpinnings the direction of the program was subject to drift for there could be no clear policy, only vague ambitions. The documents which preceded the establishment of the program suggest that both the parks branch and the advisory board envisaged compiling an inventory of historic properties and then setting about to preserve them.[1] Such an approach – involving the purchase of real estate and the upkeep of historic buildings – implied a large budget and extensive powers to appropriate designated sites. But these means were not automatically provided, nor did it appear that they would soon be forthcoming.

In order to obtain money and power the program needed an approved policy, yet in order to devise a policy it needed to know what it could expect in terms of formal authority and a budget. Without either it was largely powerless to act. Most new government programs face similar circumstances but usually quickly manage to reach firmer ground with both legislation and budgetary support. However, the men responsible for launching the government heritage agency were faced with aggravating circumstances. First, in the years following the war fiscal restraint meant that the government was reluctant to commit money to new programs, especially ones not tied to improving the economy, and it gave little encouragement that it would fund historic sites on anything but a limited scale. Second, the advisory board was itself powerless to implement policy. It was further hindered by the regional differences represented by its members and by serious conflicts of personality.

Another complicating factor lay in the dual nature of the heritage program: preservation and commemoration. The first involved the treatment of heritage structures to prevent their destruction and to facilitate their enjoyment by the public. Treatment could range from the mere stabilization of ruins to the renovation of dilapidated buildings. At a more elaborate level it could also include the restoration or replication of vanished elements or even the reproduction of the entire structure. Another part of preservation was the provision of visitor services. Access roads had to be built and possibly even an interpretive museum constructed. Commemoration, on the other hand, involved erecting monuments, usually with an explanatory inscription, at sites associated with specific historic events. A commemorative plaque could be erected at a heritage development but usually existed independently from preservation efforts.

Preservation and commemoration each had different implications. The first was obviously more expensive than the second and presented a series of technical and operational problems. Commemoration, on the other hand, was relatively inexpensive, and the problems it posed were theoretical rather than practical: what were the relevant historical facts, and how did they relate to larger themes of national development? Preservation and commemoration signified different approaches to the past. An historic park with restored buildings and an interpretive museum tended to recreate a former way of life. A commemorative inscription concentrated on relating a particular episode from the past. Commemorations, therefore, tended to be more directly didactic than preserved historic buildings which were educational in a more visual way.

Because of the different approaches to the past that preservation and commemoration each implied, they appealed to diverse segments of the heritage movement. Commemorations were most popular with heritage advocates in Québec and Ontario where nationalist history had its strongest roots. Preservation tended to be a priority in the Maritimes where individuals and local groups promoted the development of heritage sites as part of a distinctive regional landscape. These competing regional tendencies came together on the board.

The different orientations of the board and the branch also tended to fragment the development of a comprehensive program. The branch was naturally drawn to the practical problems of heritage development; the board to the more intellectual concerns of commemoration. Yet both sides needed to co-operate to implement a balanced treatment of both aspects of the heritage program.

Unfortunately, initial development within the branch and the board worked to prevent this synthesis. First, the branch was unable to organize itself properly to administer a new heritage program and lacked both the legislation and money necessary to develop a comprehensive policy. The board, dominated by the concerns of its Ontario and Québec members, concentrated more and more on commemoration. This orientation of the board meant that Maritime commemorations were largely disregarded: the board identified sites for preservation, but took little interest in their subsequent development. This left the branch, by default, to look after the problem of preservation in the Maritimes and elsewhere, yet its limited capacity to act resulted in little being done. Moreover, the branch lacked a definite policy of its own or a comprehensive overview from the board and it was prone to proceed in a desultory and *ad hoc* fashion.

THE BRANCH

By 1919 the Dominion Parks Branch of the Department of the Interior seemed well prepared to undertake a heritage policy. A survey of historic sites had been started before the war, and by 1919 the branch was in contact with numerous organizations and individuals compiling suggestions for national historic sites across the country. The branch had acquired two historic properties, Fort Anne and Fort Howe, as national parks and was in the process of acquiring others in the west. Following correspondence with Judge Savary in Annapolis Royal, Harkin formed the opinion that historic sites needed a separate definition in the Dominion Parks and Forests Act, and his staff began preparing language governing historic sites for the new national parks bill then being contemplated. In 1919 formal responsibility for historic sites in the branch rested with Assistant Commissioner F.H.H. Williamson, but by the end of 1920 a new man, A.A. Pinard, was hired to be solely responsible for historic sites. Subsequently, it became Pinard's task to cope with the array of potential sites needing assistance and to come up with a workable policy. In 1921 he replaced Williamson as secretary to the board and also endeavoured to facilitate the activities of that body.

The heritage section of the new parks bill was the first attempt by the branch to establish a firm basis for an historic sites policy. In June 1920 Harkin prepared a memorandum for the deputy minister in which he said: "At the end of this year the historical survey should be reasonably complete and it will then be possible to formulate more comprehensive and detailed plans with respect to historical sites of national interest and importance. It is quite obvious already that special legislation will be necessary if this work is to be carried on efficiently and satisfactorily. A bill has been prepared by this Branch with that object in view."[2] The bill was based on the British Ancient Monuments Act and had been drafted by the branch legal adviser with advice form board member and lawyer, James Coyne. More precisely, it had been prepared as section II of a new parks bill, "The Dominion Parks and Historic Sites Act." It would have granted sweeping powers to the branch to acquire historic property and even provided for a temporary preservation order to seize control of heritage structures that were in imminent danger of destruction.[3] But this bill never reached Parliament.

There were a number of reasons for the bill's failure. Surprisingly, the most obvious reason, that it might have contravened the British North America Act, which gave jurisdiction over property to the provinces, does not seem to have been raised as an issue. Although the bill was drafted by lawyers, neither they nor other department officials concerned with getting it passed raised the question of it being ruled *ultra vires*. Moreover, parallel attempts to have a new national parks bill enacted managed successfully to establish the principle that parks could remain a national responsibility. This was at a time when it was generally conceded in Ottawa that western natural resources were properly in the domain of the province in which they were found. As a consequence, a movement to transfer federal Crown land to the provinces culminated in the Dominion-provincial agreements of 1930 which formally transferred jurisdiction of natural resources to the provinces. Through this process, Harkin managed to convince the government that national parks were properly a federal responsibility because they benefited the population of the country as a whole, and their tourist revenue was spread over a number of regions. Thus Meighen, who was minister of the Interior in 1920, was able to speak of the national purpose of parks in his defence of the increased appropriation of the branch.

> The Dominion of Canada – not federally speaking – that is, not speaking from the standpoint of the treasury, but the people of the Dominion – derive great advantage not only in having recreation grounds for the whole Dominion, but in the tourist traffic, the returns per acre being figured by the officers of the department as being more for the rocks and waste lands of our parks than even for our wheat fields. Undoubtedly, speaking from the whole national standpoint, it is good public business to maintain these parks.[4]

The government could have extended the agreements it signed with British Columbia, Alberta, Saskatchewan, and Manitoba dealing with national parks to apply to national historic sites. It also had the precedent of the National Battlefields Act. This 1908 legislation established the National Battlefields Commission which managed the Plains of Abraham in Québec City as a national treasure. The commission soon broadened its mandate, however, to include a number of other historic sites in or near the city. And with the government willing to spend money in order to develop property for its tourist potential, why should provinces or individuals object? But the government was not eager to spend money; it wanted to cut the deficit, and only determined action at the ministerial level could extract more dollars for department activities. And more dollars is what the heritage bill would have required.

Another problem facing the enactment of historic sites legislation was that in its initial phases it was tied to the new national parks bill. Because of the protracted negotiations with the provinces, this bill was held up and not passed until 1930. Meanwhile the whole issue of historic sites was overshadowed by other larger concerns at both the branch and departmental level.

In 1919 the branch allocated a budget of five thousand dollars for historic sites. In 1920 the budget was increased to ten thousand dollars. Subsequently, it was Pinard's job to enlarge the position of historic sites within the branch. As it soon became evident that the heritage legislation would not receive quick passage, it was incumbent on Pinard to urge the department to commit itself to historic sites projects on an individual basis. Only in this way could his budget be increased sufficiently to implement a workable policy.

In embarking on this difficult task, theoretically Pinard had a number of assets. First was his own competence. While possessing no background in the heritage field, he had a number of qualifications that made him an effective government official in this area. He was one of the few francophone officers in the department and was able to develop good relations with the historical community of Québec and Sulte, the Québec representative on the board. As a former army officer, he had a good rapport with the Department of Militia and Defence, still important as the custodian of a number of historic forts, and with E.A. Cruikshank, chairman of the Historic Sites and Monuments Board and also a military officer. Both of his superiors were personally interested in his work. Harkin, the commissioner of national parks, continued to be involved with historic sites through his membership on the board. Williams, the assistant commissioner, had been instrumental in early efforts to establish the historic sites program and remained closely tied to efforts at devising a policy. While Pinard had virtually no staff of his own in these early years, he could call on help from other divisions in the branch. The engineering division, the largest component of the branch, could be expected to provide the necessary field work as well as technical advice. Following the appointment of a town planner in 1921, who subsequently formed the nucleus of the architectural division, another source of technical expertise became available. Even the migratory bird officers of the wildlife division could be called upon to inspect sites from time to time. Elsewhere in the department was the research capacity of the Natural Resources Intelligence Branch. In 1922 Ernest Voorhis prepared an *Historical Account of the Territorial Expansion of the Dominion of Canada from 1497 to 1920* for the branch. Later Voorhis would compile a survey of historic forts and trading posts of the French régime and the English fur trading companies.[5] A more regular source of historical advice for Pinard was the Public Archives of Canada whose staff responded to a number of his queries. Archaeologists and anthropologists with the newly-defined national museum were another well of technical knowledge.

In practice Pinard faced a number of difficulties. Ironically, it was the growing strength of the parks branch that contributed to the weakness of Pinard's situation. Since the war, Harkin and Williamson had been moving the branch into a number of areas parallel to park development. They had established wildlife reserves and a wildlife division to preserve vanishing species of game. Harkin, through his involvement with the Advisory Committee on Wild Life, had become concerned with arctic sovereignty. Larger issues of tourism also interested Harkin and by 1923 he had engaged a director of publicity whose office gradually embraced promoting Canadian tourism.[6] These emerging priorities within the branch diffused what should have been two of Pinard's potential sources of power. As a result of these other interests, Harkin and Williamson could not focus on historic sites. Moreover, some of the new divisions, particularly the new wildlife division under Hoyes Lloyd, had a legislative basis and were better positioned to receive a larger share of the branch budget than the historic sites division.[7]

What made matters worse was that while Harkin continued to expand the activities of his branch, he also continued to concentrate his budget on the development of western parks. This was necessary for a number of reasons. Government spending was largely determined by a program's ability to produce revenue, and the growing popularity of the Rocky Mountain parks made them a potential revenue producer.[8] But the postwar tourist trade was undergoing a revolution as the increasing popularity of the automobile changed the character of recreational travel. In order to stay competitive, the branch had to undertake an extensive program of road building. The engineering difficulties posed by the mountains made this an extremely costly venture and accounted for a large part of the branch budget throughout the 1920s. Moreover, road building was a priority of the engineering division, and during the summer months most of its staff was busy in the west, unavailable for visits to eastern historic sites. As in the case of Pinard's other allies in the branch, then, shifting priorities relegated historic sites to the back burner.

The historic sites office was further handicapped by its reliance on other divisions. While the engineering division was willing to do the historic sites field work, historic sites did not need to develop its own operational capacity. Matters were further complicated by the fact that the largest historic site, Fort Anne, as a regular national park, remained outside of Pinard's jurisdiction – its budget was separate from the historic sites allocation. For these reasons, the profile of Pinard's office remained extremely low within the branch and the department. During the fiscal year 1922–3, when expenditures of the branch totalled more than one million dollars, nearly 15 percent of the department's budget, the expenditures of the historic sites division were under twelve thousand dollars, slightly more than 1 percent of the branch total.

THE BOARD

Along with the parks branch, the other main component of the government's heritage agency was the Historic Sites and Monuments Board of Canada, created to advise the minister on the selection, commemoration, and preservation of national historic sites. From the beginning the board's function was limited to that of an advisory body. Although it could recommend the designation of potential historic sites and advise on their subsequent development, it was the responsibility of the Dominion Parks Branch of the Department of the Interior to carry out this development, to decide what was or was not feasible. Thus the branch could heed or ignore the board depending on its own assessment of the situation. This situation was partly ameliorated by Harkin who represented the branch in the board's decision making. Nonetheless, the private members of the board were a distinct group representing aspects of the heritage movement at the local, regional, and national levels and therefore presenting particular concerns that had to be dealt with by the heritage agency. At the same time, although they shared many common concerns, the private members had among them a number of contradictory perspectives. There were, then, from the outset two levels of possible conflict: the ideals and aspirations of the heritage movement had to be shaped to the capability and will of the branch to implement them, and the priorities of each board member had to compete with those of his colleagues. This organizational tension was a dominant theme of the early history of the government's heritage program.

The work of the board was, to a considerable degree, shaped by the personalities of its members. Along with Harkin, one of the guiding lights on the early board was Brig. Gen. E.A. Cruikshank (1853–1939), known as the "General." His solid reputation as an historian and his proximity to the bureaucracy in Ottawa made him an ideal chairman. He was a leading figure in the Ontario and Canadian heritage movements: he had been elected to the Royal Society in 1906 and had been a member of its preservation committee. He was a founding member and secretary of the Historic Landmarks Association and was to be president of the Ontario Historical Society from 1920–2. Retiring from his post as director of the military historical section of the Department of Militia and Defence in 1921, he continued to reside in Ottawa. Although Cruikshank suffered from ill-health, was often shy and remote, and spent his winters in Jamaica, he remained a leading figure on the board until his death.

Although he had a national reputation, Cruikshank's historical interests were largely focused on a single area, the War of 1812 in southwestern Ontario. Much of his earlier career had been spent as a gentleman farmer and part-time militia officer in the vicinity of Welland on the Niagara peninsula. In this guise he had combined an interest in the region's history with the military and undertaken a study of the role of local militia units in the War of 1812. He became active in local historical groups and was a member of the Lundy's Lane Historical Society, aiding that group in its effort to have the Lundy's Lane monument erected. Cruikshank's reputation as an historian was largely based on *The Story of Butler's Rangers and the Settlement of the Niagara Area* (1893) and a nine-volume *Documentary History of the Campaign on the Niagara Frontier in 1812–14* (1896–1908). He also published numerous pamphlets on military subjects, including the battles of Lundy's Lane and Queenston Heights.[9]

But though Cruikshank prided himself on his scientific approach to history and scorned biased nationalist histories, his writing betrayed certain implicit ideals. It is apparent, for instance, that he was an adherent of the loyalist cult, which argued for the centrality of loyalist traditions in Canadian life, and that he placed considerable importance on this tradition in explaining a series of wars against American aggression. Moreover, as an Ontario nationalist, Cruikshank saw in the past ideals which were instructive for the present. He has been quoted as saying: "All men justly owe a debt of gratitude and remembrance to those who have gone before them, particularly when they have been their benefactors in many ways, especially when blessings of moral, intellectual, social, and political freedom had been won by the endeavours of their ancestors and their kinsmen."[10] This is a key statement to understanding Cruikshank's position on the board, for it not only suggests the élitist view of history to which he subscribed, but points to his sense of mission in bringing history to the people.

Also from Ontario, James H. Coyne (1849–1942) was closely associated with Cruikshank. Born in St Thomas, Coyne had as a teenager served with the local militia during the Fenian raids of 1866. Later he attended the University of Toronto where he won a gold medal for high standing in languages. After a short spell as a teacher, he trained as a lawyer. He returned to St Thomas to practise and was eventually appointed county registrar. Although law remained his work, his love was history, and Coyne became a gifted amateur historian. His particular interests in this sphere were southern Ontario prehistory and the early exploration and settlement of the province. His writings included *The Country of the Neutrals, from Champlain to Talbot* (1895), "Indian Occupation of Southern Ontario" (1916), and a translation of a work by the Sulpician priest Dollier de Casson who, with his colleague de Galinée, explored the Lake Erie region in the seventeenth century.[11] Coyne's activities in local history led to his involvement in the larger realm of the heritage movement. In 1897 he was elected president of the Pioneer and Historical Association which he was largely responsible for re-organizing as the Ontario Historical Society. He was elected a Fellow of the Royal Society of Canada and served a term as president of its English letters and history section.

A leading member of the Ontario heritage movement, he, like Cruikshank, shared many of its attitudes. He was a friend of George Taylor Denison and shared his views about the rightful development of Canada as a nation within the British Empire. Speaking to a Victoria Jubilee Day gathering at Port Stanley in 1897, he said: "The genius of the Anglo-Celtic race is towards union, toleration, federation, righteous law and administration. Its instinct for extension of property is ... justified by its extraordinary success in governing inferior races upon principles of justice and equality."[12]

The Québec representative on the board, Benjamin Sulte (1841–1923), introduced a distinctive perspective toward national history. Sulte was born and raised in Trois Rivières where he missed many of the advantages useful to an intellectual career. Orphaned at six, he left school at ten to work in a shop and for the next few years held a variety of mentally and financially unrewarding jobs. This bleak career path changed when he joined the militia in 1863 at the time of the Trent Affair between the United States and Great Britain and became a noncommissioned officer. This experience encouraged him to pursue the military as a career, and he was accepted into the military school run by the British army in Québec. He received a commission and in 1866 saw active duty along the Canadian-United States border. About this time he began writing and after demobilization worked for a short time as a journalist. He served as an aide to George-Étienne Cartier and following Confederation worked as a translator and then as a senior official in the Department of Militia and Defence.[13]

Sulte's reputation is largely based on his poetical and historical writing. He began writing poetry in the 1860s and then became committed to chronicling the history of his people. His major work, *Histoire des Canadiens-français* (1882–4), although lacking in scholarly brilliance and at times turgid in style, was, spanning seven volumes, at least epic in scale and helped establish his reputation as an important Québec intellectual. His other historical writing covered topics ranging from the Québec militia and local history to the West and Louis Riel. This output helped Sulte gain prominence at the national level. He was a founding member of the Royal Society of Canada and was elected its president in 1902.

Sulte was a particular type of French-Canadian nationalist. His imperialism and abiding interest in the militia tied him closely to his Ontario counterparts. He translated "God Save the Queen" into French.[14] Yet he did not subscribe to the belief that it was the destiny of the Anglo-Saxon race to dominate North America. Like George-Étienne Cartier and Wilfrid Laurier he looked to Canada to bind two nations in the bosom of a single state. Thus, while maintaining imperial sentiments, he was also deeply committed to preserving the French-Canadian heritage as a fundamental part of Canadian culture. He was an original member of the *Cercle des dix,* a group of intellectuals which modelled its organization after the French Academy, and was for a time president of the Ottawa chapter of the St-Jean-Baptiste Society. He was also interested in informing English Canada about his heritage and in 1897 read a paper to the British Association of Toronto, "The Origins of the French Canadian."[15] But although Sulte joined the board as *primus inter pares,* he was, at seventy-eight, feeling the effects of his age, and his lack of vigour considerably diminished his influence.

The two members from the Maritimes were well connected to heritage groups in their respective provinces. As a Loyalist, William Odler Raymond (1853–1923), venerable archdeacon of Saint John, had much in common with his Ontario colleagues. As honourary captain in the Third New Brunswick Regiment, he was interested in the militia, and he was a Fellow of the Royal Society of Canada.[16] Unfortunately, Raymond's influence on the board was practically negligible. He never attended a meeting and lived, after 1919, in Toronto where he was being treated for the illness from which he eventually died.

William C. Milner from Nova Scotia, on the other hand, made a considerably greater impression. Milner (1846–?) was originally from Sackville, New Brunswick. He received a BS from Mount Allison University and later founded the Chignecto *Post.* Although not a prolific writer, he apparently wrote newspaper articles on historical subjects. It was probably this activity which led to his appointment as Maritime representative of the Dominion Archives.

This job enabled him to travel around the region scouting possible historic manuscripts, and in the process he became interested in the problem of protecting historic sites.[17] He penned a report in 1916 to the premier of Nova Scotia on the plight of some of his province's heritage properties and subsequently lobbied the federal government as well.[18] He had the support of his superior, Dominion Archivist Arthur Doughty, who had recommended his appointment to an expanded National Battlefields Commission in 1914.[19] Milner differed from his colleagues on the board in a number of key respects: he was interested in Maritime historic sites as resources to be preserved and developed and he was not interested in sites as expressions of local nationalism. Of all the members, then, his approach to the past was the least concerned with commemoration and the most concerned with preservation. Moreover, as a Maritimer his approach to the past differed from that of his colleagues from Québec and Ontario in being less deliberately anti-American in its nationalism.

Unfortunately, Milner's governing characteristic was his inability to work in a group situation. This tendency minimized the degree to which the particular concerns of the Maritimes, which Milner represented, affected the general orientation of the board. One candid observer who met him in 1920 stated that he seemed "to be sore about things generally, perhaps it is chronic with him, appearances point that way."[20] Master of the angry letter, he soon directed his invective at his colleagues on the board and once wrote to Williamson: "If any of the said members imagine that the Maritime Provinces is a sort of buffalo range and I am a keeper, it is not my business to undeceive him any more than it would be necessary for me to enter into controversy with any long legged, ugly mugged, yelping cur that thinks he owns the sidewalk."[21] He referred to Cruikshank and Harkin as "clerks in the Public Service, who show indications either from travel or reading of no fitness for the service."[22] Such extreme language made it very difficult for the gentlemen on the board to tolerate his company. Even Raymond, his Maritime colleague, cited Milner's difficult personality as a reason for wanting to resign.[23] So, while representing a particular regional viewpoint, Milner's effectiveness was virtually nullified because he alienated the rest of the board.[24]

While the members of the Historic Sites and Monuments Board brought individual and regional concerns into the federal heritage agency, together they formed a separate corporate identity contiguous with the historic sites branch. The board was not merely a forum for individual expression. It was an advisory body which sanctioned the creation of national historic sites. The nature of its identity, however, was ambiguous. At one level the government was willing to credit the board with the whole heritage program. On commemorative monuments and in speeches in Parliament it was the board, not the branch, which was identified with national historic sites.[25] At the same time the board had no formal powers. It depended on the branch to implement its ideas and so was subject to the same constraints of inadequate funds and legislation. And, for its part, the branch had its own objectives for a heritage program. It was the responsibility of the board as a whole to locate its position between the two poles of absolute dependence on or total independence from the branch. It had to accommodate the regional and individual perspectives of its members in a practical agenda that could be carried out by the division. The formative years of the board, then, were taken up with defining its relations with the branch and endeavouring to set objectives and implement a program over which it could exercise a modicum of control During this period the strengths and weaknesses of its members played a significant part in determining the eventual character of the board.

The particular concerns of the members emerged at the first meeting of the board held in Harkin's office in October 1919. Harkin tabled an agenda, and those present elected Cruikshank chairman. An array of objective and subjective criteria for designating national historic sites was then introduced. The principal decision made at this meeting was that only sites deemed to possess national significance should be recommended as national historic sites. This implied the existence of a graduated hierarchy of sites having local, provincial, or national historical meaning. It also assumed the existence of a finite number of sites that could be objectively determined. Deciding on the abstract principle of national significance was one thing, agreeing on its application was another and posed serious problems. Given the diverse attitudes of the members toward historic sites, such consensus proved almost impossible to achieve. Conflict was only avoided by each member confining his own standards to his particular area of responsibility.

Besides establishing a selection standard, the first meeting of the board also set out a procedure by which sites could be chosen for national designation. The members were formed into committees which would survey potential sites in each region and then propose sites to the whole board for final approval. Initially these committees were composed of Milner and Raymond for the Maritimes, Sulte for Québec, Harkin and Cruikshank for eastern Ontario, Coyne for western Ontario, Cruikshank for western Canada, and Sulte for trails and explorations of western Canada. Despite the attempt to provide regional representation, Ontario received a disproportionate amount of attention during the early years of the board's operation. The fact that two of the members were from that province coupled with Cruikshank's strong bias in favour of loyalist sites, meant that Ontario received far more designations than any of the other provinces.

Initially it was decided that the board should meet only once a year in the spring, at the time of the Ottawa meeting of the Royal Society. To help keep it functioning in the interim months an executive committee, comprised of Harkin, Cruikshank, and Sulte, was formed to attend to questions relating to the implementation of the program in Ottawa. In 1919 the committee met with the minister of Militia and Defence to discuss the transfer of heritage forts to the parks branch. And in January 1920 Cruikshank, Sulte, Harkin, and Williamson met with Maj. Ernest Fosbery, a member of the Royal Canadian Academy, to commission the design of a plaque which could be erected at all national historic sites. Membership on this committee gave Sulte and Cruikshank stronger ties to the bureaucracy and meant that their personal agendas could carry more weight than other members in affecting the department's heritage policy. As Sulte was handicapped by old age, Cruikshank had the field of executive responsibility virtually to himself. And, while enhancing Cruikshank's power, the committee diminished Milner's position as a board member by delegating Williamson to inspect and report on Maritime sites.[26]

Milner saw the committee's delegation of the work in his region to a government official as a personal affront and complained to Harkin's superior, W.W. Cory, the deputy minister of the Interior.[27] In justifying this action, Harkin pointed out that Milner had been given an opportunity (albeit an extremely limited one) to report on Maritime sites but he had not acted in time. In any case, he added, "it was never intended that the Board should have administrative functions or that the free action of the Department should be trammelled or prejudiced by any recommendations the Board might make."[28] Relations between Milner and Harkin deteriorated rapidly after this, with Milner assuming the role of a spiteful troublemaker, and Harkin studiously ignoring the Nova Scotia member. In June Milner wrote Prime Minister Borden complaining about the board and threatening to resign if better representation were not accorded the east.[29] He enlisted the support of Dominion Archivist Arthur Doughty who also wrote Borden to support Milner and reiterate his proposal that the heritage program be administered by an expanded National Battlefields Commission. Harkin countered these attacks by seeking the support of his minister, Arthur Meighen, to defend his administration and by preparing cooly reasoned rebuttals. In a confidential note to Borden's private secretary, Harkin placed the blame on Milner for the board's incapacity to function.

In passing I may say that apparently the only mistake made was in the nomination of Mr. W.C. Milner as a member. With the exception of the first meeting of the Board his presence at the meeting has been productive of nothing but delay and trouble. In fact conditions had become such that some weeks ago it became clear to me that it would be utterly impossible to retain any of the other members of the Board if Mr. Milner was to continue a member and I am recommending to the Minister that Mr. Milner be removed.[30]

Harkin won, and Milner was ignored and eventually replaced, but such a situation did little to advance the cause of Maritime historic sites or enhance the reputation of the government of Canada in eastern heritage circles.

By the second full meeting of the board, in May 1920, Cruikshank's views emphasizing commemoration in central Canada were unchallenged. Accepting the limited appropriation of the branch for historic sites work, the board agreed to scale down its objectives and concentrate on the selection and marking of national historic sites. Williamson was authorized to send a circular letter advising local heritage groups that they could not expect much help in preserving or restoring historic buildings. This read in part:

There appear to be so many sites which are eminently national in character requiring preservation, that you can doubtless readily understand if the Board were to concern itself with every site the work would immediately develop into something so large that it could not be carried out at this time of financial stringency.

For this reason the Board feels that it should not at this time attempt to do too much in the way of restoration or preservation work of sites, except to prevent such sites from deteriorating beyond repair.[31]

Here one can see the influence of the branch, through Harkin, manœvering the board to act as a shield against adverse public criticism. The board was then led to approve the policy of compiling an inventory of historic sites. Cruikshank proposed that this could result in the publication of an illustrated handbook. The board then approved Fosbery's design for a commemorative bronze marker and presented the initial recommendations for commemoration.[32] Not surprisingly, these favoured sites in Ontario and Québec. Subsequently, with the increasing debility of Sulte, the role of Cruikshank and Coyne on the board was enlarged even further. Cruikshank assumed some of the responsibility for designations in western Québec, and Coyne took charge of the survey of western sites. The late summer of 1920 found Coyne on an extended tour of northern Ontario and the western provinces during which he compiled a long list of potential designations.

The dominance of Cruikshank and Coyne, and, to a lesser extent Sulte, had several implications for the board. First, although they were interested in the larger sphere of Canadian history, their immediate concerns stemmed from their particular interests. For Cruikshank this was loyalist settlement and the War of 1812. For Coyne this was Indian settlement and French exploration in western Ontario, while for Sulte it was heroic defences in the preconquest era. Second, these men were primarily interested in commemorative plaques rather than larger developmental projects of preservation or restoration. Partly this concentration on a commemorative program reflected the reality of a minimal budget. But they were not vocal in demanding a greater appropriation or in insisting on the development of heritage buildings. This is because their own approach to the past tended to emphasize the instructive over the visual.

These men also influenced the historic sites program with their shared belief that historic sites could play a useful function in contemporary society. While Harkin stressed the commercial dimension of sites as tourist attractions, these men emphasized their moral value in helping to civilize a raw and materialist society. They considered themselves part of an educated élite whose duty it was to impart proper values of patriotism, duty, self-sacrifice, and spiritual devotion to young and new Canadians and members of the lower orders of society. As members of this cultural élite, they tended to suspect many of the trends of modern society, particularly its materialism and disregard for abstract values. The past not only evoked images of a better, premodern age, not cheapened by the tastes of popular republicanism, but also provided examples which served to instruct the present.[33] One occasionally catches glimpses of this sense of historical mission in the correspondence of the board. In 1920, for example, Coyne wrote Cruikshank to tell him of his success in organizing support for the cause in Sault Ste Marie. "Immediately after my arrival I saw his Honour Judge Stone, who is greatly interested in historical matters ... I find a great deal of enthusiasm amongst prominent citizens, from which I augur a very successful career of the new society."[34]

Writing historical inscriptions which related historical events to larger principles was, of course, an important way of carrying out this mission. Incidents from the past could be tied to larger historical principles which justified ideals implicit in contemporary society. These larger principles inevitably related to views of nation and, as they were elucidated by the Ontario representatives on the board, this nationalism took a particular form. For Cruikshank, as for Denison before him, the Loyalists and their descendants formed the basis of much of what was good in contemporary Canadian society. They not only legitimized imperialist sentiment but provided a tradition of successful resistance to American hegemony. It is not surprising, then, that many of the sites proposed by Cruikshank reflected this theme. The list of nominations which he submitted at the May 1920 meeting included proposals to commemorate the foundation of Kingston and Prescott by Loyalists, the Battle of the Windmill (1838), the Battle of Crysler's Farm (1813), Glengarry House, and the construction of the Rideau Canal. In subsequent years Cruikshank would champion the cause of many local efforts to commemorate sites connected with Loyalists and the War of 1812, so that before long there was a veritable palisade of historical markers along the St Lawrence glorifying episodes of resistance to the American invasion.

The problem with Cruikshank's approach to the past was that it was exceedingly narrow and, like the work assailed in Herbert Butterfield's *The Whig Interpretation of History,* was susceptible to fallacies inherent in studying the past for the sake of the present. Like the Whig historians, Cruikshank tended to be selective – recognizing only those incidents from the past which supported his view of the present – and reductionist – explaining the present by constant reference to events in the remote past. While supporting local efforts to promote loyalist sites, he resisted initiatives to recognize positive attributes of the rebels of 1837. They were not to be regarded as heroes, and the only site specifically associated with them until the 1930s was Montgomery's Tavern. William Lyon Mackenzie was repeatedly turned down for commemoration despite the entreaties of the Toronto historical society, the York Pioneer and Historical Society. In 1928 the secretary of the society complained to Harkin about the board's decision that Mackenzie was merely of local significance, pointing out that "there surely should be some memorial for William Lyon Mackenzie, who did more for liberty in Canada than almost any other man in our history."[35] But Mackenzie was not a national benefactor in Cruikshank's view, and his interpretation of Canadian history greatly influenced the board.

Coyne's attitude toward the past was slightly different from Cruikshank's. First, he was more interested in the possibility of preserving an historic landscape for its own particular merits. He regarded the Southwold Earthworks as being valuable more for their rarity as an example of our prehistoric heritage than as an historic lesson. He saw the spiritual as well as the pragmatic value of heritage and once argued, for example, that it was quite appropriate to "gather up and preserve the traditions and legends of the Ojibwas and Mississagas with reference to 'old forgotten far off things, and battles long ago.'"[36] Second, although he remained primarily committed to a program of commemoration for its educational objectives, he took a less defensive view of Canadian national development. He was, therefore, less concerned with marking battle sites and more interested in sites that indicated the stages of Canadian geographical expansion. Coyne's original list of proposed designations therefore included Port Dover, Southwold, Sault Ste Marie, Port Arthur, and the Midland mission site.

For Coyne the quintessential national historic site was the "Cliff Site" at Port Dover which honoured the spot where Dollier de Casson and de Galinée stood when they claimed the region for New France in 1670. The commemoration of this seemingly minor historical episode reveals much of Coyne's attitude toward the past and the function of historic sites. Coyne had a personal stake in this site. He had written a scholarly account of the expedition and translated part of the explorer's journal. It was the first site he proposed and the first one actually to be commemorated in Ontario. He managed to persuade the branch to abandon its plan of erecting a boulder cairn on the site and instead build a more expensive concrete cross, supposedly depicting the original one put up by the French explorers. The inscription which he prepared for the monument read:

Near this spot, March 23, 1670, was erected a cross with arms of France and inscription claiming sovereignty in the name of King Louis XIV over the Lake Erie region, as shown in the procès-verbal reproduced on this memorial placed here in 1922. Canada was ceded by France to Great Britain in 1763.[37]

Coyne was suggesting here a kind of titular succession from God to France to Britain and, perhaps, to Canada. Cruikshank, who shared this view, explained to Harkin: "In a sense, the act of possession in the name of Louis XIV may be said to be the beginning of the title of the British Empire, and of Canada in particular, to the Lake Erie basin."[38] Such an interpretation was not shared by Raymond who, in a rare communication, argued: "The action of the missionaries and their compatriots was to a large extent a political action and, as Ontario is now a British province, there is no sufficient reason for repeating the ceremony performed in 1669 (two centuries and a half ago) by erecting a cross and affixing thereto the arms of a foreign potentate in token of his sovereignty."[39] But with Raymond's absence from the meetings, the board could overlook this objection, and Coyne's enthusiasm prompted the designation of a second commemoration, "The Wintering Site," to the French expedition at Port Dover.

The importance of geography to Coyne's evolutionary view of history led him to promote sites at the Sault and the Lakehead on Lake Superior. Of the first site he wrote Cruikshank in 1920: "It has been suggested that on account of the great importance of the Sault in connection with discovery and exploration, not merely of the Upper Lakes but of the Rocky Mountains, Mackenzie River and Arctic Ocean an elaborate monument might be erected at some point along the canal, on which might be recorded the names of all the conspicuous characters commencing with Brulé, Nicolet, Radisson ... and down to La Verendrie, [*sic*] prominent explorers of the Canadian fur companies and of the arctic expeditions."[40] The implication was that the Sault was the gateway through which French and later British explorers had established the geographical limits of contemporary Canada. Consequently, the May 1921 meeting of the board recommended acquiring a site on the Sault Ste Marie ship canal to commemorate western explorers and the three hundredth anniversary of Brulé's discovery of the Sault in 1621.[41]

This notion of a gateway to the west was also applied to the Lakehead where the importance of Port Arthur was identified in relation to early trading routes and the commencement of the western stretch of the CPR. Later Coyne included Fort William, proposing the sites of the Kaministikwia portage and the commencement of the Grand Trunk Pacific. In a 1929 speech at the unveiling of the monument commemorating the portage Coyne tied two isolated episodes of transportation into an expression of nationhood. The portage route symbolized the importance of the birch bark canoe for the early commercial development of the country; the commencement of the Grand Trunk Pacific epitomized the development of a modern transportation system. He told another board member how he could relate these two images to a single theme. "This afforded me an opportunity to enlarge upon our national indebtedness to the canoe, for giving the French priority of discovery, enabling them to hold the territory north of the lakes and westward with their scattered trading posts, and to retard settlement west of the Ottawa to such a degree that it was practically a virgin territory that was awaiting the UE Loyalists when they were forced to leave their homes at the Revolution. That Canada is British to-day is largely due to the birch bark canoe."[42] As with the Port Dover inscription, French occupation was, in a convoluted way, responsible for the establishment of Canada as part of the British Empire.

Sulte's approach to the past was, like Cruikshank's, more preoccupied with monuments. Both views reflected the nationalist uses of history prevalent in their respective provinces. Sulte, therefore, became closely allied with Cruikshank in promoting the commemorative aspect of the heritage program. Cruikshank's initial submission to the branch for designations in eastern Ontario was compiled with the concurrence of Sulte.[43] Sulte's own first slate consisted entirely of nominations from his hometown of Trois Rivières – the birthplace of La Vérendrye, the Battle of Three Rivers (1775), and the St-Maurice Forges – and conformed to Cruikshank's idea of national historic sites. The Battle of Three Rivers, honouring a Canadian victory over American invaders, was obviously compatible with Cruikshank's loyalist battle sites. Similarly, the inscription for La Vérendrye, ending with the claim, "his explorations and those of his sons doubled the size of Canada,"[44] had broad national significance. St-Maurice Forges was important for being the site of one of Canada's first industries.

Sulte's approach to the St-Maurice Forges demonstrated his predilection for commemoration over preservation. Although a sizeable ruin remained on the site, he was not particularly concerned with its preservation or development. He wrote to Harkin in 1919: "All that can be done in our days is to clear away the heap of stones, in order to reach the foundation walls and plant a sign in the centre of the square thus uncovered."[45] The following year he told

Williamson that "it is needless to think of the ruins of the old Forge because what remains of them is really three or four heaps of broken stones."[46] Sulte concentrated instead on trying to obtain a monument which would incorporate a bas-relief design. Limited funds eventually permitted only the erection of a standard bronze tablet.

While Sulte's nationalism resembled Cruikshank's in its defensive form, it had a distinct French flavour which contrasted with Cruikshank's imperialism. Although the two could agree on the importance of post-conquest struggles against American domination, they viewed earlier battles from different sides. Their contradictory perspectives are revealed in the commemoration of episodes of the seventeenth-century French-English colonial wars. Here there was considerable ambiguity for Canadians as to who were the protagonists. For Sulte the heroes were clearly the French defending their homeland against the English aggressor. For Cruikshank, the wars were an episode leading up to the conquest of New France, but he was willing to bend this view to accommodate the perspective of his Québec colleague. By calling the invaders Americans and the defenders Canadians, he was able to bring the Québec view in line with his own. Still, this accommodation merely concealed rather than resolved opposing views toward the past.

The particularity of Sulte's nationalism is evident in his promotion in 1920 of the site at Laprairie. This became one of the most important sites in Sulte's pantheon when it was augmented by the commemoration of the Second Battle of Laprairie. Although two distinct sites, Fort Laprairie and the Second Battle of Laprairie commemorate a single historical event. In the seventeenth century the farming community of Laprairie formed an outpost of Montréal against English colonial and Indian invasion. There was never an actual fort, but a palisade had been erected around some of the houses. In 1691 the governor of Montréal, having received word of an impending invasion, dispatched a large force of soldiers and native allies to Laprairie to prevent a crossing of the river. What arrived was not an invasion force, but rather a raiding party of about 270 Albany militia and Indians led by Maj. Peter Schuyler. The attackers canoed up the Richelieu River to the rapids below Chambly then followed a trail through the forest to Laprairie. Their attack at night caught the defenders by surprise, but the superior numbers of the French company caused the attackers to retreat along the trail toward their canoes. So ended the first battle of Laprairie. Schuyler's men were not pursued, but a picket of French soldiers and Indians led by Captain Valrennes intercepted the raiding party and attempted to block their retreat along the trail. What followed was an intense encounter known as the Second Battle of Laprairie to distinguish it from the defence of the town. In then end Schuyler and his men managed to bypass the blockade and flee in their canoes.

Statements about the significance of the events at Laprairie appeared soon after the attack with the French and Anglo-Americans each producing their own version. The original American account of the battle is "Major Peter Schuyler's Journal of his Expedition to Canada,"[47] while the basis for the French interpretation is M. Bénac's "Relation des actions qu'il y a eu cette campagne entre les français et les sauvages anglais," written in September 1691.[48] F.X. Charlevoix's eighteenth-century *Histoire et description générale de la nouvelle France* adhered to Bénac's interpretation. It attributed the salvation of the colony to the defence of Laprairie and added that the double victory over the English dissuaded the invaders from making further attacks on the colony. Sulte's historical narratives followed Charlevoix's version.[49] Some English Canadians of the time also praised the heroism of both sides. W.D. Lighthall, in his poem, "The Battle of Laprairie," published in 1922, wrote: "Were those not brave old races? / Well here they still abide."[50]

But in French Canada after the First World War there was a growing feeling that a more exclusive heritage should be celebrated. Victor Morin, president of the Société historique de Montréal, wrote Williamson in 1920 complaining about what he perceived to be an English bias in the heritage program. "It seems that in the minds of some people, the history of our country has only commenced with the English Conquest, and any event previous to that is not of national importance and should not be commemorated. Contrary to that, we are of the opinion that the foundation and development of his country under the French Régime are of sufficient interest to be recorded by the erection of monuments."[51] The branch responded to this concern by sending Pinard to Montréal to interview members of the society and solicit their advice on Sulte's proposed designations and inscriptions. Subsequently, Sulte and the Montréal group worked in harmony to produce a series of commemorations. The monument to the Second Battle of Laprairie was one result of their collaboration.

Valrennes's determined defence against Schuyler's raid had particular significance to French-Canadian nationalists. In form the story resembles other episodes from the history of New France that were also revered in a quasi-religious way: the defence of the family fort against marauding Indians by Madeleine de Verchères and the defence at the Long Sault against an Iroquois raiding party intent on taking Montréal. These episodes have a common theme – the heroic defence, involving personal sacrifice, against the superior numbers of an alien invader – a theme that was relevant to twentieth-century Québec where the traditional values of Catholic, rural French Canada were being invaded by modern industrial society with its concomitant godless materialism and alien language.[52] In this context, the inscription of a memorial erected to mark the Second Battle of Laprairie in the 1890s had a contemporary significance. After giving the names of the French soldiers slain on the field of honour, it concluded, "un souvenir d'un fait d'armes entre le Français et les Sauvages chrétiens d'un côté et les Anglais et les Sauvages infidèles de l'autre, le 11 août 1691."[53] Here a whole cluster of images comes into play. On the one side there are *les anglais,* Americans and infidels; on the other there are French, *canadiens* and Christians. This was a clear message for continued struggle.

Cruikshank was supportive of the effort to commemorate the Second Battle of Laprairie because the nationalism that inspired it, conservative and defensive in form, was so similar to his own. He insisted on its designation at the 1921 meeting of the board, from which Sulte was absent, and where other members wondered whether it was not a duplication of the Fort Laprairie commemoration.[54] Cruikshank's only argument with Sulte's second inscription, which portrayed the combat as a struggle between French and English, was that he should substitute the word Canadian for French.[55] For Cruikshank, the long history of French Canada's defence against American aggression was perfectly compatible with the loyalist tradition. Even Denison had accorded an important role to French Canada in helping Canada develop as in independent British state in North America.[56] Cruikshank's ideology was, therefore, sufficiently broad to encompass French participation in the formation of the Canadian nation. Nonetheless, there were still important differences between his thinking and Québec nationalism.

PROGRAM DEVELOPMENTS TO 1923

The board was not completely preoccupied with commemoration. The May 1920 meeting recommended that steps should be taken to preserve Louisbourg, Forts Beauséjour, Gaspereau and Piziquid (Edward) in the Maritimes, Fort Churchill (Prince of Wales's) in Manitoba, and Fort Pelly in Saskatchewan.[57] But it was content to leave the details of their treatment to the department. As we have seen, the capacity of the parks branch to act in the area of preservation, however, was pathetically weak. Lacking its own expertise, it was dependent on the abilities of local entrepreneurs and other agencies within the bureaucracy. These allies had their own priorities and while they helped the cause of heritage preservation, they also influenced the way in which particular sites were developed. Local entrepreneurs, for example, wished to build up the importance of their projects as local attractions. Other divisions within the parks branch, influenced by their experience with natural parks, tended to view preservation as an exercise in landscaping. Throughout this process the influence of the board was remarkably absent, preoccupied as it was with commemorations in central Canada.

The two priorities for preservation in the Maritimes were Fort Beauséjour in New Brunswick and Louisbourg in Nova Scotia. There was a considerable amount of public pressure for the government to undertake preservation and development of these places, and the board had also signified their importance. Yet despite the attention they received, the branch was unable to implement even the most modest of plans.

Williamson, who was an engineer, visited Fort Beauséjour late in 1919 to report on its condition and propose means by which it could be preserved. He found a simple cluster of ruins in a farmer's field: "The ruins consist of a circular mound with piles of stone, representing the old walls. The old walls of a brick and stone vault are partly standing."[58] He recommended taking steps to have the land, which was still government property, transferred to the jurisdiction of the parks branch, and the property protected by a good fence. Similarly, Cruikshank visited the site in 1921, and the board subsequently advised "that the ruins should be preserved from further deterioration but that no actual restoration be carried out."[59] But the branch could not even accomplish this simple task.

Unfortunately, the branch chose badly in allocating responsibility for the work. In 1920 Harkin obtained the secondment of an army engineer, Capt. Harry J. Knight, to carry out Williamson's modest scheme for preservation. Knight was employed because of his previous enthusiasm for heritage work but during 1921 and 1922 the captain displayed a remarkable inability to achieve anything concrete. While at first he at least dispatched eloquent reports to Ottawa, eventually even these stopped coming, and the branch was reduced to sending urgent telegrams in a futile attempt to evoke a response from him. Finally the branch dispatched the regional migratory bird officer to investigate the situation. His report, ending with the remark, "I might add that my impressions of this man were unfavourable,"[60] caused Harkin to write Knight off as a loss. Meanwhile, Knight's inaction stalled the plans scheduled for 1922 and prevented the granting of an appropriation for the following year's work. By 1923, therefore, all that had been accomplished was the surveying of the land and its transfer to the parks branch.

Even less was accomplished at Louisbourg. Williamson also inspected this property in 1919 and recommended that steps be taken to survey and acquire the land. The following year a thousand dollars was allocated, and captain Knight detailed to supervise the work. But Knight was even less inclined to work in Cape Breton than in New Brunswick, and nothing was accomplished. Meanwhile, public criticism of the inactivity of the parks branch in the Maritimes began to mount. Partly this was stirred by Milner who used his newspaper connections to publicize the follies of the branch. Another thorn was J.S. McLennan at Louisbourg whose influence in Ottawa had increased since his appointment to the Senate in 1916. By 1922 the senator had become very critical of the branch's efforts to preserve the site and suggested to a number of well-placed people that the development of Louisbourg should be assigned to the jurisdiction of a separate commission.

The parks branch fared only slightly better in Québec where the two forts, Chambly and Lennox, were already operating as heritage attractions. At Fort Chambly, the honourary curator, J.O. Dion, had drawn up a development plan and obtained the help of the Department of Public Works. The outbreak of war suspended these plans, and Dion died in 1916, but L.J.N. Blanchet undertook to carry on the curatorial work, and it was expected that the Department of Militia and Defence would continue the arrangement after the war. As late as February 1920 even Benjamin Sulte considered this to be the ideal way to manage the fort.[61] But the military no longer considered the care of obsolete fortifications to be part of its mandate and sought a means of extricating itself from its commitment to Fort Chambly. The officer commanding the military district, under whose jurisdiction the fort lay, recommended that no more funds be spent on restoration, that Blanchet be discharged, and that the fort be razed and replaced with a suitable commemorative monument.[62] But Blanchet did not intend to lose his position without a fight and enlisted the support of contacts in nationalist organizations and political allies to mount a spirited defence of the old fort.

At this juncture the Historic Sites and Monuments Board intervened. Sulte recommended the preservation of the fort as a national historic site. A committee of the board met with the deputy minister of the department, Eugene Fiset, to discuss the transfer of a number of old forts, including Fort Chambly, to the department. Seeing an honourable path of retreat, the military approved the transfer and provided fifteen hundred dollars from its own budget to finance urgently needed repairs. Fort Chambly was declared a national historic site, and in January 1921 was place by order-in-council "under the control of the Department of the Interior to be administered and maintained by the Historic Sites and Monuments Board."[63]

Of course the board had little to do with the fort other than to prepare an inscription for the plaque erected there. Responsibility for its development was, in fact, assumed by the staff of the parks branch, although Blanchet retained a great deal of unofficial power. In December 1920 A.A. Pinard had visited the fort and suggested the need for "a comprehensive scheme for restoration and preservation work."[64] Again in 1922 the site was inspected by an officer from the town planning division who likewise recommended a long-term plan.[65] But in these dark days memoranda and words were all that the branch could afford.

Like Fort Chambly, Fort Lennox, located on Île-aux-Noix up the Richelieu River, had been a heritage development since before the war. R.V. Naylor, who ran a cruise boat along the river between Lake Champlain and St Jean, developed the site as a stopover for his excursionists. He leased the abandoned fort and surrounding land from the Department of Militia and Defence, from whom he also obtained funds for renovation. But although Naylor's business catering to the leisure needs of American tourists was initially prosperous, his enterprise declined in the years following the war, and by 1920 the military was receiving complaints about the run down condition of the fort. Faced with reassuming direct control of the property – an option that was contrary to the department's policy of not investing money in nonstrategic sites – the Department of Militia and Defence followed up a suggestion made by the Historic Sites and Monuments Board that it transfer the property to the parks branch. As with Fort Chambly, this suggestion was readily acceded to, and the order-in-council transferring Île-aux-Noix to the jurisdiction of the parks branch was formally approved in May 1921.[66]

As with Fort Chambly, the Historic Sites and Monuments Board had very little to do with the development of the site. Pinard set about negotiating for the acquisition of the museum collection put together by Naylor, and parks officials visited the island to devise a development plan. It is clear that from the beginning the parks branch considered Île-aux-Noix as much a recreational area like the Thousand Islands National Park as an historic site. Although Pinard considered enlarging Naylor's museum (which had been housed in the former officers' quarters), he also suggested that "a very welcome feature of the parks work, which is popular on the St. Lawrence River should be inaugurated, namely: permission to camp on the island, even within the fort boundaries."[67] As a result, the initial proposal for developing the site called for the creation of a bird sanctuary, the clearing of campsites, and the establishment of picnic grounds and a bathing area. Even Cruikshank, who visited the site on behalf of the branch, was enthusiastic about this concept and noted that the island could "easily be converted into a magnificent park."[68] But here again there was no money, and, apart from hiring a caretaker, the branch undertook little work on the site until after 1923.

The branch embarked on a number of initiatives in Ontario without success. It acted on Coyne's recommendation to acquire the site of the Southwold Earthworks, but the farmer whose land it was on asked what was considered to be an unreasonable price, and the board agreed that it would be better to await a new parks act which would grant powers of expropriation.[69] Another priority was Fort Henry in Kingston which had been identified as an important military remain. The army was approached and seemed sympathetic to proposals to preserve it but insisted that it needed the fort as an ammunition depot.[70] The two sites of a seventeenth-century Jesuit mission in Ontario associated with the martyrs Brébeuf and his brethren, Ste Marie I and Ste Marie II, had likewise been recommended for preservation and development, but here too the branch met with failure. The owner of Ste Marie I, near Midland, was not eager to part with the land, although he was amenable to allowing a commemorative plaque to be erected.[71] The site of Ste Marie II, on nearby Christian Island, which was administered by a local Indian band, presented better prospects. Pinard visited the site in 1920 and drew up a development plan calling for the excavation of the vanished palisade and the erection of a museum.[72] But although the band was agreeable to this proposal, the government was not, probably because the land could not be ceded to the branch. By 1923, therefore, the parks branch had not acquired any historic properties in Ontario for restoration purposes.

The branch had better luck in Manitoba where it was able to

acquire the former important fur trade depot, Prince of Wales's Fort, in a relatively simple fashion. The fort, situated on Hudson Bay near the mouth of the Churchill River, was among the first sites recommended for designation as a national historic site, and even Cruikshank was in favour of its acquisition by the parks branch.[73] Although the fort had been partially destroyed by the French following Samuel Hearne's capitulation in 1782, it remained one of the oldest extant forts in the country. And though far from any large settlement, the publication of J.B. Tyrell's edition of Hearne's *Journey to the Northern Ocean* by the Champlain Society in 1911 awoke considerable interest in the old fort. This interest was enhanced following the establishment of Churchill as the northern terminus of the Hudson Bay Railway. Being already on Crown land, it was a relatively simple matter for the parks branch to acquire the necessary fifty acres of land in 1920. But there was no settlement at Churchill at that time, and the branch and the board agreed that there was, therefore, no need to proceed with a development scheme.[74] The local RCMP constable was delegated to keep an eye on the property. He kept the branch informed of its steadily deteriorating condition and periodically unearthed some of the old cannons.

Further west there were few obvious resources suitable for development into heritage parks. One exception was Fort Pelly in northern Saskatchewan. As early as 1913 the parks branch had sought to reserve this property from settlement, believing it to be the scene of the first Northwest Territorial Council (a quasi-colonial body operating when the area was under the control of the Hudson's Bay Company) held in 1877. Subsequently 960 acres were ceded by the Hudson's Bay Company to the Department of the Interior.[75] But because there was some confusion about whether it was the right place, and few were sure if anything remained of the fort, the project was eventually abandoned. Elsewhere in the west little was happening. The branch communicated with the Hudson's Bay Company about acquiring the well-preserved Lower Fort Garry near Winnipeg but lost interest when it learned that the site was leased to a local automobile club.[76] In his first list of proposed designations, Coyne had suggested the commemoration of Fort Langley, on the lower mainland of British Columbia. But although one crumbling building remained of the old fur trade post, the branch did not even take steps to commemorate the site.

While the branch concentrated on preservation, the board was left to direct the work of commemoration. But in this area, too, the achievement was quite dismal. Partly this reflected the uneven composition of the board, for during the period 1919 to 1923 only Cruikshank, Coyne and, to a limited extent Sulte, laboured away at their appointed tasks. Of a total of thirty-seven sites actually designated, only three were from Nova Scotia and New Brunswick, five from Manitoba, and none from Alberta and British Columbia, while eleven were from Québec and eighteen were from Ontario. The branch aggravated this lop-sided situation by its slowness to act upon the board's recommendations. The first plaque was not put up until 1922 and this, at Port Dover, was the only one erected that year. Only twenty-four plaques appeared in the following year, and of these none was in the western provinces, while a total of fifteen were put up in Ontario.

The lacklustre record of the government's heritage program attracted widespread criticism. Public opinion in the Maritimes complained about the inability of the government to do anything at Louisbourg or Fort Beauséjour. Questions were asked in the House of Commons about the lack of preservation being undertaken in Ontario.[77] Letters were received complaining about the neglect of western sites. One Alberta pioneer, who had previously advised Harkin on western sites, wrote in 1923: "the reason why I can take no further interest in the 'Historic Sites and Monuments Board of Canada,' or the 'Historic Landmarks Association of Canada,' is because neither of them is 'of Canada,' but only of Ontario, and other Eastern Provinces: they are practically foreign bodies as far as my old province Alberta, and BC are concerned."[78] But Harkin already knew that reform was necessary.

Harkin realized the deficiencies of the Historic Sites and Monuments Board back in 1920 but held off taking action, hoping that the passage of a historic sites bill would provide a legislative basis for a newly constituted board.[79] Likewise he expected that a definite policy approved by cabinet would justify an expanded branch. But neither the legislation nor the policy was forthcoming, and in 1923 events had reached a crisis point calling for some kind of immediate action. Harkin persuaded his minister to appoint a new advisory board by order-in-council, thereby signifying a new beginning.

CHAPTER THREE

Fresh Start: The Search for a Policy Continued, 1923–30

By the spring of 1923 Harkin and the National Parks Branch were ready to begin anew the task of implementing a federal heritage policy. Key to this fresh start was a newly constituted board which convened for the first time in Harkin's office in May of that year. While it included the leading figures from the old board – Cruikshank, who was re-elected chairman, Coyne and Sulte – the querulous Milner had been dropped, and new people added who, it was hoped, would strengthen representation outside central Canada. These were J. Plimsoll Edwards from Halifax, Nova Scotia, J. Clarence Webster from Shediac, New Brunswick, and Frederic Howay from New Westminster, British Columbia. The appointment of Webster and Edwards promised a new deal for historic sites in the Maritimes. Harkin wrote both men a letter of welcome to the board in which he said: "I regret that the Maritime provinces have been somewhat neglected in this respect, owing to certain conditions, which I do not wish to recite at present, but now that representation is included in the personnel of the Board, it is hoped that favourable progress will be made."[1] Similar high hopes were pinned to Howay who was expected to represent not just his native province but the entire region west of Ontario.

Concurrent with the reformation of the board, the national parks branch was reorganized to better facilitate the administration of an historic sites program. Pinard's office was elevated to a division within the branch, and G.W. Bryan was appointed his assistant. Engineers from the engineering division were seconded to particular historic sites to oversee their development, and the town planning division was asked to report on the long-term planning of sites. Harkin also managed to double his historic sites budget for the fiscal year 1923–4.

Harkin pressed both the board and the branch to try to resolve their previous difficulties. He recommended that the board broaden the themes of its commemorations to include social and economic history and he rationed the monuments in the four regions – the Maritimes, Québec, Ontario, and the west – to five each per year. At the same time the branch continued its attempts to draft an acceptable piece of heritage legislation and redoubled its effort to gain control of historic properties for future development. Harkin, meanwhile, tried again to obtain a definite commitment from the minister for a heritage policy and an adequate budget to carry out its aims.

Reorganization of the branch and board alleviated many of the problems encountered earlier. The branch was better able to supervise historic sites under its control, and the board included stronger representation from the east and west. But these improvements only served to illuminate a more profound problem: there was still no policy. Without this and an accompanying operational budget, there was no clear commitment from the government for a comprehensive approach to historic sites. The larger scale planning needed to properly organize a representative heritage program was therefore impossible to achieve. Getting such a commitment would form one of Harkin's principal objectives during the 1920s, yet his requests fell on deaf ears. As a result the parks branch proceeded on an *ad hoc* basis. Sites were acquired where the local lobby was strongest, as in the Maritimes, or where there was least difficulty, as in the case of military sites which were already government property. The result was that the preservation side of the program, which was the activity most closely identified with the branch, became disproportionally loaded with military sites from the Maritimes.

The agency was further hampered by a lack of legislation defining the roles of the board and the branch. Without a clear mandate, the board was discouraged from actively participating in larger preservation issues. Influenced by the bleak budgetary forecasts, and by Cruikshank's own predilection for commemoration, the board came to focus almost exclusively on that side of the program, usually responding to questions of preservation only at the direction of Harkin. The Maritime members, meanwhile, who were primarily interested in preservation, began dealing directly with the department instead of acting through the board. The result was a growing fissure in the agency that would eventually weaken both the branch and the board.

Despite the addition of stronger representation from the east and west, the board continued to be marked by structural imbalance. Ontario had twice the representation of the other regions, while the provinces to the west had only one member. Representation from Québec remained weak, and the views of that province were barely represented on the board. With Cruikshank as chairman, military sites continued to receive a great deal of attention, particularly in Ontario. The greatest difficulty facing the board, however, was that it was still unable to resolve the problem that national significance meant different things to different regions. Moreover, the Maritime members felt that the board was dominated by a central Canadian bias, while in the west, Ontario and Québec nationalisms came into conflict over the significance of events of the 1885 North-West Rebellion.

THE BRANCH

A major priority for the branch in this period was to attain control of a series of large but obsolete fortifications – the citadels at Halifax and Québec City, Fort Henry in Kingston, Ontario, and Fort Rodd Hill in Esquimalt, British Columbia – and develop them as historic attractions. Added to Fort Anne and Louisbourg in the Maritimes, forts Chambly and Lennox in Québec and Prince of Wales's Fort in Manitoba, these would have formed a highly visible chain of historic parks across the country. To acquire these the branch needed a definite policy from the government and a vastly increased budget to develop them. It was unable to achieve either of these objectives.

Harkin enlisted the support of the board in his effort to obtain a larger appropriation. In 1925 it passed a resolution stating that "the present appropriation for carrying out its work is quite inadequate, and that the importance of the preservation and commemoration of historic sites and the consequent stimulation of interest in our national history requires a greatly increased grant."[2] Not receiving a reply to this plea, the members pursued it the following year by petitioning the minister in person. But this attempt also failed to elicit a positive response.

Although individual members seemed to be lukewarm to the military sites targeted by the branch for development, the board gave its collective support, again in the form of a resolution: "That it is desirable that the fortications of Quebec and other military posts should be preserved owing to their historic interest and the fact they are an attraction to travellers and the general public, and that their preservation is a stimulus to the growth of a healthy national and patriotic sentiment in our land, and that unless some early action be taken it is apparent that these structures will become dilapidated and fall into ruin."[3] At the same time the board expressed doubt about whether the government of Canada should assume total responsibility for developing these sites. Appended to its resolution was the recommendation that "it is the opinion of this board that special commissions should be appointed with powers to take over and preserve such fortifications and other structures, and that the Dominion, Province and City concerned should contribute in due proportion to the upkeep thereof."[4] Harkin agreed with this position to a limited extent. He was convinced that provincial and municipal governments should contribute to the expenses of potentially lucrative historic properties. But it is also evident that he did not agree with the concept of local commissions, fearing that they would compromise the autonomy of the federal government in general and the branch in particular.[5] Confused in this way about the best means to proceed with the development of large heritage projects, both the branch and the board lost the opportunity to participate in significant developments in the following decade. The restoration of Fort Henry in the 1930s, for example, undertaken through a cost-sharing agreement between the federal government and the province of Ontario, was carried out independently from the operations of the parks branch.

Meanwhile, Harkin had been carrying on his own campaign to obtain a larger commitment from the government. In 1926 he prepared a memorandum for the deputy minister presenting his case for a heritage policy, a policy that would require a significant increase in the historic sites budget.

> Personally I feel that this Department has taken the initiative in this important work and has unquestionably the sympathy and backing of public opinion that we should continue with it and expand our activities for the needs of the case. I would suggest that a definite sum of say $150,000 be made available each year which in about ten years should restore all the national historic structures to good condition, and that the sum required each year for their maintenance thereafter would be small.[6]

But no decision was forthcoming, and by 1929 Harkin expressed his disappointment in a frank memorandum to the deputy minister, W.W. Cory. "I am afraid that for a good many years the only policy that has been in effect in regard to [these sites] has been one of passing the buck. It seems to me the situation should be definitely faced and a decision reached that either the Government should or should not take action in the matter of preservation."[7] But by this time Harkin must have realized that the chances of his branch getting a mandate to undertake the development of the citadels, Fort Henry, and Fort Rodd Hill, were slim indeed. In 1928 he had advised his staff, busy trying to calculate estimates for the ensuing year that: "probably we can get best results by dealing with individual cases."[8]

Dealing with individual cases meant concentrating on sites that were already within the system, or where the branch had a commitment for future development. But even within this limited sphere it was hamstrung by the lack of a policy. Throughout the early 1920s the branch still hoped to receive new legislation which would distinguish between national parks and historic sites. A bill drafted in 1924 defined an historic site as "any monument, structure, building, relic, remain or fossil, the preservation of which the Historic Sites and Monuments Board of Canada considers a matter of national interest."[9] While this legislation was pending, the branch seemed reluctant to undertake large developments under the authority of the old act. In the meantime, however, it had a commitment to property already under its control and faced mounting pressure from Maritime interests to establish Louisbourg and Fort Beauséjour as national parks like Fort Anne. While circumstances forced the branch to concentrate on the Maritimes, it also sought to extend its program of preservation in Québec, Ontario, and the west. With larger resources and a better organization than before, it was able to make some significant progress in these areas. But lacking commitment from above or consistent direction from the board, its activities in this sphere were characterized by frustration and confusion.

While the preservation of the ruins at Louisbourg remained a priority for the branch, it gradually realized the enormous problems which this task posed. In 1923 a heritage enthusiast from Sydney, Nova Scotia, Walter Crowe, described to Harkin the legal tangle which ensnared the property. An act of the Nova Scotia legislature was necessary before the government could acquire the central portion of the property, still under the formal jurisdiction of Kennelly's defunct development corporation. A further seventy acres had been sold to a railway company and was then under the control of the Department of Railways and Canals. Other small holdings on the site had been acquired by local residents who had built clapboard houses and grazed their sheep among the ruins. All of these rights – those of the Kennelly Trust, the railway right of way, and the small landholdings – had to be alienated before the site could be developed. The first of these difficulties was soon overcome. With Crowe's assistance, the branch convinced the Nova Scotia government to pass an act in 1924 allowing the transfer of the Kennelly property to the government of Canada, bringing the surviving ruins at last under its jurisdiction.[10] At the same time the Department of Railways and Canals agreed to transfer its right of way to the jurisdiction of the parks branch. The remaining property, requiring money to obtain, was not immediately accessible.

Still, by 1924 the parks branch had sufficient property to undertake a heritage development at Louisbourg. The question then arose of how to proceed. In 1923 the town planning adviser to the branch, the prestigious British planner, Thomas Adams, had visited the site and recommended lines along which it could be developed. Adams's approach to historic sites was essentially visual. He regarded heritage property as something to be preserved in an aesthetically pleasing setting[11] and influenced the future direction the town planning division would take. "Many of these [historic] sites ... even if preserved, might easily lose the national significance that should adhere to them if they were surrounded by ugly and depressing buildings and were destitute of pleasing and impressive approaches."[12] These views reflected Adams's approach to Louisbourg. "The site of the Fort is an impressive one, apart from its exceptional historic interest. There is a certain grandeur and wildness about the harbour of Louisbourg and the surrounding hills, as seen from the site, that makes one feel in a mood to enjoy its romantic character and visualize the historic events that were witnessed from it." Adams "enjoyed witnessing everything on the site – whether in ruins or changed by nature – that belonged to the distant past, and disliked the structures or improvements that had been carried on in recent years."[13] Consequently, he recommended minimal intervention on the site, such as removing some of Kennelly's additions and landscaping to enhance its natural beauty.

Unfortunately, in 1924 Harkin was unable to proceed with even these modest proposals. He was still awaiting funds and heritage legislation that would enable the development of Louisbourg as something other than a national park. Possibly, too, he was aware of just what a sinkhole for expenditures the remote Louisbourg was. In 1926 he wrote with specific reference to the old fortress that "for some years it has been increasingly difficult to secure appropriation especially if the proposition involves actions which call for annual expenditure practically for all time to come."[14] But in postponing action the branch lost the initiative, and when the site did come to be developed it was with the involvement of powerful local forces.

Operating independently of both the branch and the board, and Pinard's division was the curator of Fort Anne in Annapolis Royal, L.M. Fortier. As the fort had been created a national park in 1917, it received a separate budget from that of historic sites, and Fortier reported to Harkin, not Pinard. His considerable energy made him a third force for heritage development in the Maritimes. He ran Fort Anne as his personal fiefdom, and it was Fortier and not the branch who took the credit for its preservation. In this independent role he was often a thorn in the side of the branch. Webster described him in a pamphlet on Maritime sites, written just before he joined the board, using misinformation Harkin suspected as coming from Fortier.

> That the old fort is now, in a small way, beginning to come into its own is due in part to the recent Tercentenary which has been celebrated there, but, in the main, to the efforts of Mr. L.M. Fortier, who has, after years of work in Ottawa, settled in Annapolis Royal, giving his whole time to the care of the fort and to the development of a historical museum in the officers' barracks. He has had no salary, insufficient and inefficient assistance, a meagre grant doled out with plenty of obstructive criticism by the Parks Commission of the Department of the Interior, in whose care the fort has for some unknown reason been placed.[15]

Among the obstacles the department later placed in Fortier's way was a formal reprimand from Harkin for allowing the unauthorized erection of plaques in the park which included ones put up by the Canadian Society of Technical Agriculturalists and the Canadian Pharmaceutical Association.[16] He was also chided for allowing his museum collection to expand beyond local history. Nevertheless, despite this friction Fortier became an important influence on historic site development as the fort and museum became a model for other sites.

An example of Fortier's influence on the historic sites program was his involvement in the commemoration of and initial plan for the development of the site of the stronghold built by Sieur de Monts, Champlain, and their companions in 1605. The site of the habitation was then just a farmer's field near Lower Granville, a few miles east of Annapolis Royal. It was Fortier who named the site and first proposed it as a national historic site in 1922.

"There is a site here that ought to be marked, and I wish very much that you kindly bear it in mind. I mean the site of the first French fort, or 'habitation' of Port Royal."[17] Enthusiasm for this site stemmed from the recently published Biggar edition of Champlain's works issued by the Champlain Society. Consequently, it was also among the first proposed for designation by J. Plimsoll Edwards following his appointment to the board.[18] Fortier's Annapolis Royal Historical Association provided a draft inscription for the plaque and, although Edwards and Webster were deputized to write the plaque text, the final version followed the lines of the one submitted by the local group.

The site received prompt attention from the parks branch, and in a gathering organized by the Annapolis Royal Historical Association and attended by a brass band, a cairn and plaque were unveiled in August 1924. Fortier's group continued to take a paternal interest in the site. It disapproved of the original caretaker appointed by the department and after a short campaign managed to assume responsibility for the care of the cairn and small plot of land.[19] The local association soon turned it into a model site: its members kept a finely groomed lawn surrounded by a post and chain fence and erected a flag pole from which flew an immaculate Red Ensign.

Fortier should have felt justified in resting on his laurels won at the Lower Granville site. But in 1927 Fortier became involved in a scheme to develop the site into something far different from a grassy field with a stone cairn and bronze plaque. That year he met Harriet Taber Richardson, a New Englander, who, along with others from her country, spent her summers at Lower Granville. No doubt influenced by the detailed descriptions in Champlain's narrative and encouraged by the example of colonial Williamsburg, Richardson and Fortier decided that a replica of Champlain's habitation on the original site was not only feasible but a worthy heritage project.[20] Together the two were able to organize considerable support for their idea. The historical artist, C.W. Jefferys, drew a facsimile of the original building, which was widely reproduced. Richardson formed an American affiliate of the Annapolis Royal Historical Association which promised to raise ten thousand dollars for the reconstruction of Champlain's buildings. With this support behind him, Fortier wrote the minister of the Interior, Charles Stewart, in November 1928 outlining his proposal and requesting a grant to match that promised by the American group, and a commitment that the department maintain the development.[21]

In the ensuing discussion between Stewart and his officials in the department, Harkin and his staff were surprisingly warm to Fortier's proposal. Harkin outlined the reasons for his support in a memorandum to the deputy minister: "I may say that when Mr. Fortier outlined the proposal to me it seemed to me that it was one that not only would have a very strong appeal to the French population of Canada but also have a very strong appeal to a very large number of people throughout the continent."[22] Although at this time Harkin professed not to want to justify historic sites in terms of their value as revenue producers, he could not help pointing out "that this old Fort reconstructed, and its story as the cradle of literature on the North American continent properly exploited, could be made a real shrine for literary and would-be literary people and that, of course, means tourist dollars."[23] It is noteworthy that both the local enthusiasts and the parks branch wanted to promote the site for its North American, not its Canadian significance. Such an approach would have appealed to American tourists, but it is also a comment on the region's approach to history.

In spite of Harkin's rare support for this heritage reconstruction, Stewart decided against it on grounds of cost, and the matter was dropped. The advent of the depression which served to distract the American supporters meant that the issue was not raised again for a number of years.

The branch fared better in Québec where it already had two sites under its control more or less functioning as historic attractions. In 1927 it obtained a special appropriation of five thousand dollars to carry out much needed shoring up of the river bank at Chambly whose erosion threatened to undermine the walls of the fort. At Fort Lennox on Île-aux-Noix there was no curator as at Fort Chambly to act as guide. There the historic sites division was left to initiate its own interpretive scheme. This had three aspects: historical plaques, a pamphlet, "Guide to Fort Lennox," and the museum. The Historic Sites and Monuments Board had commemorated both the fort and a naval engagement from the War of 1812, and two plaques were erected at the entrance in 1927. The inscription for the Fort Lennox tablet suggested some of the historical significance of the island beyond that of the fort:

> A gateway to Canada and advance post against Iroquois and other invaders. Island fortified by the French before 1759, additional works by the Americans in 1775. The whole place rebuilt by the Imperial Authorities during the period from 1812 to 1827.[24]

The pamphlet, first issued in 1923 and consisting of twenty-seven pages, explained some of the military history of the island and guided the visitor around the fort and other points of interest such as the cemeteries and the site of the former dry dock.

The museum was the most obvious means of interpretation and yet it was ill-adapted to illustrating the island's history. Although given space in the refurbished interior of the officers' quarters and provided with a showcase and identifying labels, the exhibit, which was based on the collection of the former proprietor, was badly out of focus. The artifacts in the display – buttons, swords, and muskets – were not there to explain life at the fort but were shown for their inherent antiquity. Drawn from different periods and places, the collection did not even illuminate an episode or a period of history but rather existed as a jumble of assorted goods. This situation was aggravated when the branch accepted for display items donated by the War Trophies Commission, including German helmets and rifles, an antitank gun, six machine guns, and one aeroplane.[25] Subsequently, the military donated forty surplus rifles complete with bayonets and scabbards.

The erratic collection was the result of the lack of a clear objective for the museum. Gradually one emerged, and it is possible to infer from the casual suggestions for the operation of the museum that the historic sites division had in mind more of a local history museum than an interpretive display. A letter to the caretaker, written over Harkin's signature in 1924 but drafted by Pinard, attached a photograph of the interior of one of the rooms of the Fort Anne Museum as a guide for the exhibit at Fort Lennox. The letter also provided advice on extending the collection: "You will also note in this photograph a number of handicraft articles, no doubt manufactured by the inhabitants of the locality in Nova Scotia, and the Department is very anxious, if at all possible, to secure any of the handicraft work which has been manufactured by the inhabitants of the locality of Fort Lennox, for the purpose of the museum."[26] The division also endeavoured to acquire artifacts associated with the history of the region and sent a car load of memorabilia supposedly connected with the 1838 Battle of Odelltown where rebels were defeated in their attempt to liberate Canadian provinces from British control.

During this same period the parks branch finally managed to acquire heritage property in Ontario for preservation and development. Fort Wellington, in Prescott, consisting of a wooden blockhouse, caretaker's residence, and two other minor buildings surrounded by an earth and palisade fence, was acquired from the Department of Militia and Defence in 1924. Although long outdated as a military structure, it had been kept in a fair state of repair by local militia units. Only minor repairs were necessary, therefore, to make it a functional historic park. In 1925 a museum was created and filled with the usual eclectic souvenirs. Situated across the river from New York State, it soon became a popular tourist attraction.[27]

Elsewhere in Ontario there were few projects for the preservation of historic buildings. Partly this reflected the attitude of Cruikshank who concentrated on simply commemorating historic sites and usually disliked large scale heritage developments. In the 1920s his was still a prevalent attitude among members of the Ontario heritage movement. Nevertheless, the historic sites division took some small initiatives during this period. A martello tower in Kingston was repaired, and the branch acquired Fort St Joseph on St Joseph's Island near Sault Ste Marie.[28] It repointed the surviving masonry and cleaned up the site, although little else was done to make it operational as a heritage site until a later period. The branch achieved another small objective in the region in 1929 when the property containing the Southwold Earthworks was acquired for twenty-five hundred dollars.[29] It was still a rarity for the branch to acquire private property, and only Coyne's determination made this acquisition possible. Unfortunately, there were meager resources here to be developed. Preliminary archaeology uncovered few extant remains, and the branch merely erected a fence around the grassy mounds.

West of Ontario even less heritage development was carried out during the 1920s. Prince of Wales's Fort at Churchill, Manitoba remained in limbo awaiting completion of the Hudson Bay Railway. In British Columbia the combined efforts of the Native Sons of British Columbia and a local preservation committee helped the branch to acquire the old fur trade post of Fort Langley on the lower mainland in 1925.[30] Initially only the standard plaque was erected, but in 1927 the division took some minimal steps to preserve the last remaining building. Logs were chinked, new stairs built, and wooden floors laid.

THE BOARD

The members of the board influenced the direction of the heritage program in two different ways, collectively and as individuals. As a unit the board assumed responsibility for commemorations by selecting national historic sites and writing inscriptions for the plaques, while giving moral support and occasional guidance to the historic sites division in its preservation work. The new members gave it a slightly different orientation. Although Cruikshank, the most experienced member, was reinstated as chairman and given additional power through his participation with Harkin on the executive committee, the influence of his lieutenants, Sulte and Coyne, had dwindled considerably. Sulte's handwriting at this time reflected his frail condition and he was to die at the end of the year. Coyne was seventy-four and, although he remained on the board until 1930, his participation in this period was limited. Two of the new members, on the other hand, J. Clarence Webster from New Brunswick and Frederic Howay from British Columbia, were full of vigour and rising stars in the heritage movement. J. Plimsoll Edwards, who represented Nova Scotia from 1923–5, and his successor, Walter Crowe, from Sydney, were distinctly subordinate to Webster. Although they influenced the selection of sites in their province, it was Webster who represented the Maritimes on the board, just as Howay spoke for the west. Webster and Howay joined Cruikshank, then, as leading members of the board.

While the branch concentrated its energy on devising an historic sites policy and implementing a wider program of preservation, Harkin as the branch representative on the board also looked to improving the program of commemorations. Shortly after the formation of the new board, therefore, he sent a letter to each member recommending that "more attention be given to the social and industrial sites associated with the early history of Canada, and that we must deviate from the military sites more than in the past."[32] The board, while agreeing in principle to this ideal, had difficulty making it operational. Coyne noted that "the practical difficulty is in distinguishing between sites of national, provincial and merely local importance."[33] While a national railway obviously met the criteria of national significance, on what basis could an educational institution or a local industry be justified?

In spite of Coyne's apprehension and Cruikshank's continued predilection for things military, the board made a concerted effort to comply with Harkin's request. Commemoration of "first things" became one way to justify national significance. In 1923 Halifax had been awarded a designation for the site of the first printing press in Canada.[34] The 1925 meeting of the board elaborated on this principle, recommending the commemoration of the sites of the first railroad, steamship, paper mill, salt works, petroleum well, and electric telegraph.[35] The members found other worthy topics. Howay proposed marking some pioneer mining ventures in British

Columbia and Alberta,[36] Cruikshank, all government canals,[37] and Webster, the site of the first coal mined for export in Canada.[38] Traditional topics of the fur trade and exploration further augmented this new direction.

While the board on the whole was not particularly interested in preservation, it was willing to support departmental initiatives to acquire military sites. This not only accorded with the interests of the Maritime members, but also with the outlook of Cruikshank. Military sites were also an obvious choice for preservation. Other categories of buildings were not easily recognizable for their national significance and military property was usually easily acquired as it was already government owned.

In 1924 the board passed a resolution which said that "the preservation of existing martello towers and blockhouses is justified from a national standpoint in view of their significance as a type of military architecture, and that descriptive inscriptions be prepared by the Chairman and Mr. Harkin, plaques to be placed on all such martello towers and blockhouses now existing within the jurisdiction of the Department and elsewhere when permission can be obtained."[39] At the same time the board, including Harkin, deliberately excluded other categories of historic architecture. Similarly, it would not consider any church or graveyard as a national historic site.[40] The reasoning behind this policy seems to have been a desire not to become swamped with requests to which the branch could not accede.[41] But the effect was that all of the properties acquired in this period for preservation were military sites.

The Maritime point of view was much better represented during this period than it had been before 1923 and it reflected far better than other regions, except Ontario, the aspirations of a regional heritage movement. J. Plimsoll Edwards, from Halifax, was past president of the Nova Scotia Historical Society and a minor but respected writer on historical subjects. Still, there were limitations to his representation. Edwards, it seems, was not attuned to his region's concern with preservation, placing what was considered to be inappropriate emphasis on commemoration. An opposition MP from Nova Scotia rose in Parliament to say, "I know Dr. Edwards very well indeed; he is one of the most important officials of the Nova Scotia Historical Society. But the way they proceed is to put a brass tablet on some old building and let it go to pieces."[42] Although he seems to have maintained amicable relations with the rest of the board and proposed a number of sites in the Halifax area, he resigned in 1925. He was replaced by

Walter Crowe from Sydney, a local judge and newspaper publisher, who was interested in Cape Breton history generally and the preservation of Louisbourg in particular.

But it was Clarence Webster's abilities that raised the profile of the Maritimes in the national heritage program. Until the summer of 1920 Webster (1863–1950) had been engaged in a distinguished career as a physician and surgeon. Born in New Brunswick, he had studied medicine at Edinburgh and practised in Montréal before becoming chief of obstetrics at a large Chicago hospital. He enjoyed an international reputation in his field and published a book on obstetrics that was a standard text for many years. Forced to resign because of ill health, Webster retired to Shediac where he channeled his great energy into the collection of Canadiana and research on local history. Being a man of means and education, he traveled regularly to England and France as well as to the archives in Ottawa in pursuit of artifacts and documentary evidence. In this way he became acquainted with many experts in the Canadian community including Dominion Archivist Arthur Doughty.[43]

Webster contined to develop his interests and research as a member of the heritage movement following his appointment to the board, publishing a number of books and pamphlets on local history. He became active in both the Royal Society of Canada and the Canadian Historical Association and was elected president of the latter in 1932. He also became keenly interested in the cultural development of the Maritimes. His book, *The Distressed Maritimes* (1926), attempted to link poor economic achievement with moral and cultural backwardness. He was instrumental in the establishment of both the Nova Scotia Archives and the New Brunswick Museum, contributing parts of his substantial collection to both of these institutions. A member of what has been termed the Maritime Rights Movement, a loose group concerned with improving the situation of the Maritime provinces in Confederation, Webster was named to the 1931 royal commission to investigate railway conditions in Canada. As a result, by the 1930s at least, he had established a number of important political connections.

Webster possessed plenty of confidence and was an aggressive member of the board. He made the conditions of his continued participation clear at the beginning when he wrote the minister of the Interior stipulating that his co-operation with the federal program depended on more attention being paid to historic sites in the east. "There is very great dis-satisfaction [*sic*] in these parts with regard to the lack of attention paid to historic sites and monuments. Senator McLennan assures me that you will do your best to correct the short-comings of your predecessors and it is to be on the strength of such assurances that I have accepted your invitation to join your Commission."[44] With Webster on the board the Maritimes had a strong voice, and much of the preservation undertaken in both New Brunswick and Nova Scotia reflected his influence.

Perhaps the most important lacuna in Webster's representation was in regard to Prince Edward Island. Back in 1920 the board had opined that the province did not merit any national historic sites.[45] Webster, who undertook to represent Prince Edward Island along with New Brunswick, was inclined to share this view. In 1925 he informed Harkin that, "of course there is very little of national importance in the development of the Island and it would be foolish to appoint a member."[46]

Webster shared with his colleagues the belief that commemorations served a useful patriotic function in educating citizens about common traditions. Although Webster did not have the same nationalist mission as his colleagues from Ontario and Québec, and, having lived in the United States for many years did not share their anti-American sentiments, he did share with them the conservative impulses which informed the commemorations of historic sites with something like a religious mission. Webster had strong old world beliefs. He enjoyed being in the company of titled people and traveled regularly to cultural events abroad. At the same time he distrusted modern values with their strong emphasis on material success, physical gratification, and the rule of the common man. Conscious of belonging to a cultural élite, he felt it his duty to instil proper values in the masses by preaching the importance of the arts, music, and history for, as he explained, "it is because there is no demand for these higher pleasures and civilization, and because the desire of the flesh overpowers the needs of the spirit, that the people prefer automobiles to culture."[47] Cruikshank, imbued in the conservative ideology of loyalism, and Coyne, whose bedtime reading included books in classical Greek, would have concurred with this sentiment.

It was as a "custodian of culture," then, that Webster participated in the commemorative work of the board. Years later he commented on this aspect of the board's work as one way in which civilization was brought to the Maritimes.

I have always considered the work of the Board as an educational influence of the first magnitude. The historical renaissance in the Maritimes coincides with the activities of the Board. Our many ceremonies have directed public attention to our history through the press and radio ... Thus we have exerted a marked influence in awakening an indifferent and uninformed people that they live in a country that is more than a mass of farms, factories, towns and villages, which supply the material means of subsistence for them and nothing more.[48]

It was through this sense of mission that Webster found genuine fellowship on the board.

Neither Webster's nature nor his constituency permitted totally harmonious relations in the process of selecting sites for commemoration. He possessed some of the Maritimers' paranoia that Ottawa was dominated by the views of Ontario and Québec and so was suspicious of the board's attitude to the eastern provinces. He shared with Walter Crowe the feeling that the just proposals of the Maritimes were being ignored, while relatively trivial sites in central Canada received undeserved attention. Judge Crowe wrote Webster commiserating on the failure of the board to approve of one of their proposals: "To be frank I am not greatly impressed with the breadth of view of some of our Upper Canadian colleagues. I think the time must soon come when some ginger must be infused into our discussions. I recently went over some of their projects – they are not national in the sense I take out of that word. But if they are to persist in their interpretation then they must not invoke the broad meaning of the word against our Provincial projects."[49]

Crowe tried to force a resolution of this issue of criteria at the next board meeting. In asking Harkin to place the definition of national importance on the agenda, he wrote: "I have not reviewed all the projects approved by the board, but I have examined enough to convince me that it is apparently an elastic term."[50] At the meeting Crowe stated that he "was of the opinion that all the events associated with the early struggles between the French and English were of national importance."[51] The issue was not resolved and it was raised again at the next year's meeting. There it was recorded that "Dr. Coyne was of the opinion that every fort, English or French, was of national importance, likewise the site of every battle. Judge Crowe strongly supported Dr. Coyne in this respect."[52] But the board could not accept such broad criteria, and the meeting concluded that "each site should be dealt with on its merits."[53] Such subjective criteria still favoured Ontario sites whose representatives were closest to Ottawa.

But subjective criteria also favoured those who were willing and able to outshout the opposition, and here Webster had the advantage for his enthusiasm and determination were able to overcome many of the board's inherent biases. To his other colleagues, however, it was apparent that Webster had his own biases. While continuing to recommend sites from across New Brunswick, he came to concentrate on designations near his home-town.[54] He recommended a cluster of sites to enhance the reputation of Fort Beauséjour and obtained Harkin's permission to erect smaller commemorative plaques on the site. This was viewed as parochial by Cruikshank who commented on Webster's narrow outlook in a letter to Howay in 1929: "I concur entirely with your agenda remarks as to the 'small stuff from Nova Scotia' and you might have added the still smaller stuff from New Brunswick, and I have merely suggested that aforesaid proposals should be placed on the agenda for consideration, but then there is the danger of a flare up."[55] Despite this resistance, Webster managed to triumph in a number of his pet proposals.

Unfortunately for Harkin and his staff, Webster was not always on the side of the bureaucracy and throughout the 1920s demonstrated considerable mistrust of the parks branch as the rightful agency to be undertaking heritage conservation. Few argued the cause of Maritime preservation more vociferously than Dr Webster. "It is all very well to preserve the buffalo," he wrote Harkin in 1924, "but when our historic sites suffer on their account the people are justified in raising a storm."[56] While not a troublemaker in the mold of Milner, he nonetheless could stir up trouble if he thought that the branch was dragging its feet and on a number of occasions went behind the commissioner's back to receive ministerial approval for his pet projects.

One such occasion arose soon after he joined the board. Prior to his acceptance he had met Sen. J.S. McLennan, the expert on Louisbourg, and the two had formed an alliance. Ideally, Webster favoured the establishment of a separate commission to oversee development of the site, but following his appointment to the board, he joined Edwards on a subcommittee charged with overseeing the work on Maritime sites. Subsequently, the two Maritime members met to draft a series of recommendations proposing that Louisbourg be developed as a national park. They suggested that the fishermen's buildings be removed, that more land be acquired, a caretaker appointed, and an engineer hired to mark out the original position of the streets, fortifications, and other points of interest.[57] At the same time they drafted inscriptions for a number of plaques to be erected at various places on the property. But

Harkin did not see the subcommittee as being actively involved and ignored its advice. In 1926 he described its function as being merely "to formulate ways and means toward interesting public opinion for the purpose of securing an appropriation for the future development of Louisbourg."[58] Ignored in this way and with activity in 1925 at a standstill, Webster undertook a campaign to circumvent Harkin and the branch.

Through the latter 1920s Webster joined McLennan and Crowe in lobbying for the development of the fortress of Louisbourg as a major historic site. The campaign conducted by McLennan and Webster had two phases. First they lobbied for money and then they fought to gain control over how it would be spent. The first phase began in the summer of 1927 when Senator McLennan invited members of Parliament to visit Louisbourg. As a result, in the next session of parliament L.H.N. Bourassa and others spoke in its favour.[59] The fact that these speeches were made during a debate on the renovation of the Québec Citadel placed increased pressure on the government to extend its bounty further east. Then in 1928, when the National Battlefields Bill – principally a money bill requiring seventy-five thousand dollars for ten years for the National Battlefield Commission's work at Québec City – was being debated in the Senate, McLennan raised the objection, for which he received some support, that Louisbourg deserved as much consideration as the Plains of Abraham.[60] Meanwhile, Webster did his part to mobilize public support. In December 1928 he addressed the Military Institute in Halifax on the subject of Louisbourg and while in that city "enlisted the support of several influential Liberals in bringing pressure to bear on the Government in regard to this project."[61] This phase of the war succeeded in getting the government to allocate twenty-three thousand dollars for the development of Louisbourg in 1929.[62]

With this victory behind them, McLennan and Webster were determined that they and not the branch would control how the property was developed. This was not only because of the branch's foot-dragging with regard to Louisbourg, but two misadventures at the site blamed on the negligence of the branch. The plaques, whose inscriptions had been prepared by Edwards and Webster, had allegedly not been situated with sufficient care, and some locals charged that they were inaccurate. Even though Webster had been involved in their preparation, the branch was blamed. A more serious incident occurred when the Department of Marine and Fisheries dismantled the ruins of an ancient French lighthouse on some adjacent property, provoking the ire of Webster and the chagrin of Harkin. The Maritime preservationists questioned the ability of the parks branch to care properly for historic sites and again called for the establishment of a commission of local experts to administer Louisbourg. For a brief period Webster and Crowe were mutinous members of the Historic Sites and Monuments Board, and Webster threatened to resign unless an independent commission was formed. Otherwise, he said "the development will be left to clerks in the Parks Commission, who will make mistakes, as they have already done, and will produce a thing of 'shreds and patches,' calculated to call forth only contemptuous criticism."[63]

Webster petitioned the senior Liberal minister for Nova Scotia, J.L. Ralston, to establish a Louisbourg commission. This effort produced results for in the spring of 1929 the government was openly considering the problem, and Stewart, the minister of the Interior, announced in the House of Commons that "with respect to Louisbourg and some of the other historic sites, the government hopes either to enlarge the work of the Historic Sites and Monuments Board or to create a new board to take over that work in its entirety and give it more authority to deal with the restoration of these historic sites in eastern Canada."[64]

At this juncture we need to turn the footlights up a bit to illuminate the illusive Mr Stewart before he leaves centre stage. For Stewart, as minister of the Interior from 1921 to 1930 except for the short break during Meighen's government, was central to the whole struggle of the board and the branch to define a heritage program. His importance stemmed from the fact that it was he who was expected to introduce the necessary legislation and fight for an increased appropriation in cabinet. And it was he who could quite arbitrarily, it seems, make pronouncements in Parliament on proposed re-organization of the administration of historic sites. That he took no apparent action in these areas explains why he remains so much in the background. Yet he still continued as the focal point toward which the board and the branch directed much of their energy. The Honourable Charles Stewart therefore emerges as a large question mark. Just what was he doing and thinking during this drama involving the branch and the board?

Without private papers or a history of the Department of the Interior no definitive answer can be provided, although some inferences can be made. Stewart had been premier of Alberta before joining the King cabinet in 1921 to represent an important western faction of the Liberal party. The Department of the Interior was no longer the important department it had been under Sifton: the creation of the western provinces, and the transfer of responsibilities to other departments had curbed its importance to western development. And Stewart does not seem to have been one of King's brightest ministers.[65] He does not appear to have been aware of the workings of his department, particularly of the parks branch which had grown considerably in relative importance since the end of the war. Whereas Harkin in his time had direct access to Meighen and regularly sent memoranda to Borden's office, Harkin was separated from Stewart by a deputy and assistant deputy minister who controlled the day to day affairs of the department.

Stewart seemed especially ignorant of the historic sites program, believing that it was run by the Historic Sites and Monuments Board and not the branch. His failure to understand the situation explains his odd statement in the House of Commons. He had taken Harkin's suggestion that an intergovernmental committee be placed in charge of the citadels and Fort Henry[66] and combined it with Webster's call for a Louisbourg commission. It must have surprised him when his department officials opposed this plan with Harkin pointing out that it would not be subject to government control and could possibly commit the government to embarrassingly large expenditures.[67] Stewart, then, while being potentially important to our little drama, was not really a part of it and remained like some *deus ex machina* suspended above the stage.

The real participants in the drama over the heritage program were the members of the board, along with local enthusiasts like Senator McLennan from Louisbourg, and the branch. Behind the struggle to control development lay opposing philosophies of preservation. The branch at this time adhered to Ruskin's dictum that restoration "means the most total destruction which a building can suffer."[68] This belief held that it was artificial and ahistorical to restore a building back to a former state. This was the view of the town planner, Thomas Adams, who argued in the case of Louisbourg for the preservation of its ruins, not their destruction through rebuilding. It was stated as a branch policy back in 1920 by Harkin: "If there is nothing but a pile of stones, it is not considered good policy to erect a fort on the lines of the original one."[69] Perhaps not coincidentally, this approach was also relatively inexpensive.

Webster and McLennan took quite the opposite view. McLennan had proposed the reconstruction of Louisbourg back in 1908, and through the 1920s both men argued for at least a partial restoration of the old fortress.[70] Their attitude reflected a growing trend toward historic sites in the twentieth century. With the expansion of tourist travel there was a demand for the development of historic sites as tourist attractions. Ruins were not enough, tourists wanted living museums. At the same time the development of better archival collections and technical expertise made it feasible for vanished buildings to be replicated. The leading examples of this new approach to preservation were in the United States. In 1907, using only archival and archaeological evidence and the advice of an English architect, Stephen Pell began the reconstruction of Fort Ticonderoga on Lake Champlain in New York state.[71] By the First World War it had become a major tourist attraction, and Pell spent the rest of his life creating and enlarging its museum of American history. Similar projects were undertaken elsewhere leading up to the decision by the Carnegie foundation in the 1920s to fund the reconstruction of colonial Williamsburg. These American events inspired the Maritime preservationists. Webster once said that his visit to Fort Ticonderoga was one of the happiest moments of his life.[72] Senator McLennan, who had promoted the reconstruction of Louisbourg more as a hypothetical possibility than as a practical development, began to argue for it as a viable project following the news of Williamsburg. Crowe also seemed to favour reconstruction, although he advocated more practical goals than McLennan. The real stumbling block was money. Reconstruction called for large amounts of it, and in Canada the government was the only reliable sponsor for this kind of heritage development.

These two approaches to the treatment of Louisbourg had been kept apart because the branch had become used to proceeding on its own and largely ignored the board on matters of preservation. The frustration of Webster and Crowe increased as they realized that they were being kept on the sidelines and that Harkin did not tolerate interference from civilians. As we have seen, the Maritime subcommittee had been viewed by Harkin more as an auxiliary body aiding in public relations, and in 1929 Crowe told Webster that this was probably how the branch regarded the board as well. "I have written Ottawa to ascertain just what our status is but if we are advisory I think it is by courtesy only."[73] It was this frustration at having their ideas ignored that largely prompted the mutiny of Webster and Crowe.

For his part, Harkin objected to relinquishing control of the development of Louisbourg because he feared that the branch would become committed to the costly and unrealistic scheme of McLennan. He informed his superiors of this in his memorandum rejecting the proposed creation of the Louisbourg commission: "However there is one great danger in regard to a committee of this kind and that is that it may make such extensive – and probably some impossible – recommendations so that in the end the government is only embarrassed by the work of the Committee."[74] The experience of the National Battlefields Commission, which at this time was suspected of being fiscally irresponsible or at least extravagant and was also making a large submission to Parliament, gave credence to Harkin's argument. So, persuaded by the bureaucracy, Stewart abandoned plans for a separate commission and left the parks branch in charge of Louisbourg.

But Webster and Crowe continued to fight for the establishment of a local commission to supervise the development of Louisbourg.[75] This was debated at the board meetings until 1930 when Harkin, Cruikshank, and Howay steadfastly refused to abrogate formal responsibility to nonboard members. Eventually a kind of compromise was reached where Webster and Crowe comprised a subcommittee of the board to advise on Louisbourg and they, in turn, consulted an unofficial committee made up of Senator McLennan, his daughter Katherine, the mayor of Louisbourg, and a local churchman.[76] With the lines of communication between the two sides improved, it was then possible to arrive at some compromise over the development plan.

McLennan, Webster, and Crowe had visited the site in 1928 and drew up a practical plan for the development. The plan avoided mentioning reconstruction and instead focused on limited restoration of extant ruins.[77] Crowe recommended reconstructing part of the King's Bastion and the West Gate and building a museum which would contain a scale model of the town. Crowe seems to have persuaded Webster and McLennan that this was the most that was practically possible. He wrote Harkin that "there will be those who advocate the restoration of the old Fort, the building of a museum in old French Chateau style etc. All this would cost more money than is likely to be appropriated and if too large a scheme is proposed and adopted the heart break of the delays will tire every-one out."[78] But in another communication to Harkin he pointed out that some reconstruction was necessary for "there was nothing to see and no person in the park to explain anything."[79]

Crowe's compromise proved acceptable to both sides, and in 1929 the branch moved to develop the site along the lines he proposed. A parks engineer, S.O. Roberts, was assigned to the project and he spent the summer season supervising preliminary work. With a budget of only three thousand dollars for the year, Roberts could not accomplish a great deal beyond stabilizing the deterioration of the casemates and landscaping the grounds. More was achieved in the following seasons with a budget of eight thousand dollars for 1930 and seven thousand dollars for 1931. Roberts's report for this latter year reveals something of his approach.

> The Department of the Interior is in possession of copies of the original plans of the fortress as drawn up by the famous French military engineer, Vauban, and with their aid, the engineers in charge of the restoration work are able to locate, by instruments, the exact positions of many of the former gates, walls and bastions. Several of the old streets within the fort have already been located with the aid of these plans, and a bridge has been thrown over the site of the old moat at the east gate, giving access to the old French and English cemeteries.[80]

This primitive archaeology unearthed many treasures and revealed considerable amounts of masonry. These finds encouraged some limited restoration. Although the engineer reiterated the policy of not undertaking a reconstruction, he proposed "to restore a little at a time, the principal gates, bastions and other prominent works, to indicate sufficiently the former strength and extent of the place at its greatest period in history."[81] The number of artifacts found at the site created a desperate need for a museum, and one of the houses expropriated for the development was spared demolition and converted to this end. A second house was saved as a caretaker's residence. So, although the branch remained in charge, it had moved a long way to accommodate the views of the local preservationists.

While Webster played a leading role in prodding the branch to take action at Louisbourg, he singlehandedly led the campaign to have Fort Beauséjour taken over as a national park. Although a board plaque was erected in 1923, lack of funds prevented the division from undertaking development. Local interest in the site was re-awakened, however, and an optimistic Amherst Board of Trade called for the excavation of the casemates, or bomb-proof mess, and the restoration of the magazine.[82] Webster quickly assumed the leadership of the local lobby and, as with Louisbourg, co-ordinated a powerful campaign. He enlisted the support of A.B. Copp, the secretary of state and MP for Westmorland, New Brunswick, and H.J. Logan, MP for Amherst. With these and other political friends Webster was able to pressure the branch from above to have the property developed as a national park. He made his superior position quite clear to the commissioner when he made his formal request to the branch. "I wish you to consider seriously my proposal to make Beausejour a National Park. The demand is strong. I have discussed the matter with Mr. Copp and he is anxious for it. He has asked me to go to Ottawa this autumn to see him and Mr. Stewart together. Senator Robinson of Moncton, a strong supporter of the Government, will advocate the creation."[83]

Suitably subdued by Webster's confident assertions, Harkin moved to comply with his request. He dispatched his chief engineer to report on the possible development of the site and requested a special appropriation of two thousand dollars for the following year. In 1925 he submitted a memorandum to cabinet asking for an order-in-council to establish Fort Cumberland National Park. With Webster's high-level support these requests got immediate attention, although the order-in-council was delayed for a year because Copp demanded that the park be named Fort Beauséjour, favouring the French rather than the British name of the fort. With national park status and a special appropriation largely due to Webster's campaign, Harkin was prepared to give the doctor considerable say in how Fort Beauséjour should be developed. "As it was, I think, largely through your persistent endeavours that this has finally been accomplished," he wrote Webster, "I am taking the liberty of communicating with you requesting that you be good enough to furnish me with an expression of your views as to what improvement or development work should be carried out immediately consistent, of course, with the funds which will be available."[84]

With this invitation Webster joined what would today be described as the planning team, and he and the engineer outlined the work to be done that season. Although limited by the two thousand dollar budget, it was they who established the priorities for development. The first year's plan called for the clearing of Moncton's trenches (then thought to be siege lines but subsequently found to be defensive ditches), the purchase of old cannons, building new fences, marking points of interest with interpretive signs, building roads, and demolishing the old magazine.[85] Surprisingly, it was Webster's idea to have the magazine removed. "I hold that either the magazine should be rebuilt (a proposal which would be a hideous waste of money, and which I oppose with all my might), or it should be left with only the old base or foundation."[86] The initial phase of the development, however, was largely one of engineering and construction. The grounds were landscaped, trees removed, and roads and footpaths laid out. The casemates were cleaned out, toilet and picnic facilities constructed, and the whole area made to resemble the current ideal of an historic park. Throughout this process Webster was not merely a consultant but took personal charge as if it were his own property that was being improved.

These victories at Louisbourg and Fort Beauséjour made Webster the department's consultant on preservation in the entire region. Harkin canvassed his opinion on the proposal to acquire the Halifax Citadel, to which Webster "objected seriously, stating that there were no historical associations connected therewith."[87] He was more positive about sites closer to home and in 1928 became involved in negotiations to purchase the site of Fort Gaspereau near Moncton, New Brunswick. The following year he "superintended the carrying out of various improvements there, including the removal of many bushes from the level ground," and clearing the moat of debris.[88]

With Sulte as the Québec member, the French-Canadian viewpoint was allowed to languish. He was absent from the first meeting of the new board, and it was a few months before a replacement was found after his death. The appointment of Montréal notary, Victor Morin (1865–1960), in 1924 therefore must have come as a jolt to the board. Like Sulte he had good credentials for representing the Québec heritage movement: he was a writer of Québec history, president of the Montréal Antiquarian and Numismatic Society, and Fellow of the Royal Society of Canada. But he was only fifty-nine, and his nationalism was more exclusively French Canadian than that of his predecessor. It was Morin who had written the branch in 1920 arguing for better representation of French-Canadian history in the selection of national historic sites. Subsequently, he argued for the representation of this viewpoint outside of Québec. This argument was manifest in his plea for the erection of bilingual plaques at all national historic sites.

In a letter to Harkin shortly after he joined the board, Morin suggested that, as French Canadians did not always remain at home in Québec, but traveled elsewhere in Canada, they should have the right to read federal government signs in their own language. Bilingual texts, he argued, should reflect a policy of biculturalism. "As long as we will create a separation between the two languages, granting one a certain portion of the country, and the other in some other portion, we will not succeed in bringing together the two elements of our population but we will, on the contrary, tend to keep them apart. I think therefore that not only from the point of view of justice, but principally by patriotism, our inscriptions should be in both languages everywhere."[89] Morin's proposal was placed on the agenda for the June 1924 meeting where it was discussed at length. But it was unanimously rejected.[90]

The board's decision isolated Morin, and he resigned before the next annual meeting. He was replaced by Aegidius Fauteux, another Montréal historian with strong views, but he too resigned within the year. Next came Judge Philip Demers, and then in 1930 Maréchal Nantel. But these men too seemed unable to work with their colleagues and exerted only minimal influence on the deliberations of the board. Not only were they exclusively concerned with French Canada, they were easily offended by the prejudices of their new colleagues. Howay reflected a typical English-Canadian attitude toward the nationalism of the Québec members in a letter to Cruikshank in 1935. "It is so difficult to retain a Quebec member, and every one we have had except dear old Mr. Sulte appears to have no knowledge of his own, but to go to priests and canons, and church dignitaries for information and in consequence the thing gets either a religious or a personal flavour."[91] Confronted with views like this it is little wonder that the Québec representatives did not feel at home on the board. With this lack of continuity in representation from Québec, it was left to Cruikshank, with help from Pinard, to initiate many of the designations in the province.

Other than Québec, regional representation was well articulated on the board. Webster from New Brunswick and Cruikshank from Ontario represented important points of view in the heritage movement. There was yet another forceful representative appointed to the board in 1923 who further emphasized its regional character. This was Judge Frederic William Howay (1867–1943), noted British Columbia amateur historian. Coyne had remarked on his expertise while on his western tour in 1920, and there was little doubt that the judge was the province's leading historian. In collaboration with E.O.S. Scholefield he had written a two volume history of British Columbia that remained a standard work on the subject until the appearance of Margaret Ormsby's book in 1958. His particular interests lay in the pre-Confederation period, especially early exploration and the Pacific fur trade. At various points in his life he was president of the British Columbia Historical Association, the Canadian Historical Association, and the Champlain Society, and was elected to the Royal Society of Canada.[92] Although an amateur, he took these pursuits extremely seriously, usually spending two months each summer traveling to meetings and distant archives. As a result, he was a leading member of the Canadian historical community.

Quiet and serious in manner, he too could express himself forcefully in discussion and had well-formed opinions on many historical topics. On the board Howay represented the four western provinces until 1938 and then British Columbia and Alberta until his death in 1943. Howay brought an entirely new perspective to the board because, although he was obviously an historian of the Whig school, his view of the development of British Columbia had little to do with the expansion of central Canada. He articulated this perspective in his presidential speech to the first annual meeting of the British Columbia Historical Association in 1923. "I wish on this occasion to trace in a general way and as interestingly as possible the earliest days of our Province, to strive to show that we had a story before the days of the Caribou and its wonderous gold wealth – yes, that we had a story before the foot of the Hudson's Bay trader or Nor'West trader ever trod our soil."[93] This point of view was soon translated into national historic site designations, and by the summer of 1924 a memorial tablet commemorating the early navigators was unveiled at Friendly Cove, Vancouver Island.

While building up the representation of British Columbia history, Howay helped curb Cruikshank's enthusiasm for the wholesale commemoration of sites associated with the War of 1812 and twice moved at meetings that more commemorations of that subject be suspended until the significance of the sites could be further evaluated. On occasion Howay could also circumvent the board and the branch to obtain recognition of a British Columbia site. One such case was the Great Fraser Midden, an important archaeological site in Vancouver that was in danger of being alienated by land development. In 1932 he wrote H.H. Stevens, then the ranking western member of the cabinet, to gain his support in influencing the branch: "I would be greatly pleased if you be kind enough to write to Mr. Harkin or call him up over the phone and express your interest in the matter and the pleasure it would give you personally, to have a memorial of our Board erected to commemorate this repository of ancient Indian culture."[94] But this was a rare case of political interference.

On the whole Howay was happy to work with, rather than against, Cruikshank and Harkin and became one of the most attentive members of the board. Like his colleagues, he considered himself a "custodian of culture" and endorsed the overall aims of the commemorative program.[95] He seemed more conscious than

Webster of the difference between national and local historical significance in proposing national historic sites and strove to adhere to national criteria. A characteristic Howay site in British Columbia was that in Vancouver honouring the establishment of British dominion over the region and the passing of Spanish power from the Pacific coast. Having pointed out the significance of Captain Vancouver's voyage, Howay's inscription ends with the words: "It was the dawn for Britain but twilight for Spain."[96] Other Howay commemorations focused on the effort of fur trade explorers to establish British dominion over the area. Mackenzie's rock near Bella Coola was one of his first recommendations. So, while Howay's designations had a distinct regional viewpoint, he consciously strove to avoid the parochial narrowness which he believed characterized national historic sites elsewhere in the country.

Howay agreed with Cruikshank and Harkin in that he concentrated on getting the plaques up. Unlike Webster, he was not particularly interested in preservation and therefore was not inclined to contest branch interference. He seemed quite content with the limited amount of preservation undertaken in his province and let the branch proceed with its own plans for the development of Fort Langley.

Howay was further tied to Cruikshank by their long association in the national heritage movement. Cruikshank, Coyne, and Howay had served together on the nominating committee of the Royal Society of Canada. Howay's friendship with Cruikshank deepened during the 1920s and each year Howay would generally stay at the Cruikshank's home in Ottawa while attending the meetings of the Royal Society and the Historic Sites and Monuments Board. Usually he would try to arrive at Ottawa a day or two in advance so that issues could be discussed, and a private agenda agreed upon.

Well-connected with the heritage movement at the local and national levels and attuned to the objectives of the board and the branch, Howay would seem the ideal board member. Yet his situation was not ideal for his constituency covered a great deal more than British Columbia, and it was with the designation of sites in the other western provinces that difficulties arose. Howay brought enormous energy to his task, traveling relentlessly by public conveyance, for he did not drive, and carrying on a prodigious correspondence. He canvassed academics in the history departments of the provincial universities and local historical societies for suggestions and endeavoured to be scrupulously fair in his recommendations. Yet it simply was not possible for him to represent such a vast region. Working on such a scale he was bound to overlook important points of view and, operating from afar, his minor prejudices became exaggerated. He could not tolerate A.S. Morton, for example, who advised on the commemoration of sites in Saskatchewan, and ignored his suggestions to mark sites in the southern sector of the province. Although heritage groups in Alberta regularly proposed the preservation of Indian buffalo pounds, Howay took the view that he did not wish to commemorate something associated with the virtual extinction of the buffalo. Similarly, he refused to recommend the designation of the popular site of Fort Whoop-Up, Alberta, disapproving of its association with American whiskey traders. All the while he was far more tolerant of proposals from his native province. Thus, in proposing ten sites, from which five were to be selected for commemoration in 1924, half were from British Columbia.[97]

Because of the weakness of Howay's representation east of British Columbia, the outlook of the entire board was flawed. Given the inherent tendency of the members to emphasize the history of their own neighbourhoods, it is not surprising that the prairie provinces received only a small proportion of national historic sites in the years leading up to the Second World War. Moreover, the sites that were designated were likely to be interpreted from perspectives alien to regional historical traditions. This is just what did happen in the case of the North-West Rebellion sites in Saskatchewan which embroiled the board in some of its most virulent controversies. It was a difficult episode for the board, for not only did it result in unfavourable publicity, but it forced the members to face the possible conflict between the historical and ideological significance of a site. Usually the historical events associated with a potential site involved the board in a discourse of subjective interpretation. In the case of the Cut Knife Hill and Batoche commemorations, discussions became polemics.[98]

Despite the many difficulties which they encountered on the way, both the branch and the board had established a comfortable and workable organization for dealing with national historic sites. The branch had abandoned many of its larger ambitions for acquiring large heritage projects and come to focus on the immediate needs of developments within the system. In this regard it not only recycled the expertise of its parks engineers, but accepted the direction of local advocates, men like Fortier at Fort Anne, Webster at Fort Beauséjour, and Blanchet at Fort Chambly. Gradually the views of branch officials and local enthusiasts merged to form a general ideal of a national historic park. While economic stringency still prevented largescale reconstruction, the branch seemed willing to make limited attempts to build up the historic attraction of a heritage property. At the same time it approached development of historic property in much the same way as natural parks. Access roads had to be constructed, landscaping done, and picnic sites and washrooms built. One new feature was a museum, but here the tendency was to establish a regional collection of artifacts rather than a centre for the interpretation of the site.

Collectively, the board learned to live with the branch. As a whole it left problems of preservation to the branch, although individual members were free to participate in preservation, like Webster, if they wished. Generally the board and its members approved of the branch's approach to site development. In the Maritimes, where interest in preservation was strongest, there was a tendency to promote heritage property on the basis of its general antiquity rather than for any particular historical meaning, an attitude that accorded well with the approach of the parks branch. Meanwhile, the board concentrated on its commemorative program. While there was occasional grumbling about the parochialism of Webster, or the French member, or the monomania of Cruikshank, on the whole the board was content to let each member designate sites in his particular region. Although there were gaps in representation – in Prince Edward Island, Québec and the west, excluding British Columbia – the board seemed pleased with itself.

CHAPTER FOUR

Stagnation, 1931–6

In 1930 the entire historic sites program was stalled. The National Parks Act passed that year had failed to provide the specific mandate for which members of the the board and branch had long hoped. Government spending, which had not been lavish during the 1920s, was cut to the bone at the onset of the depression. As a result, it became difficult for the branch to erect even a modest quota of commemorative plaques, and the future of its preservation work was uncertain. The members of the Historic Sites and Monuments Board seemed to lack the will to fight through these difficulties. Many of the old hands, including Cruikshank, Howay, and Webster, felt that the important work of the board was behind them anyway and contemplated disbanding.

This lethargy was partly offset by the appointment of new members who were part of a new generation of professional historians just emerging in Canadian intellectual life, and whose fresh approach gave the board new vigour. Still, there were problems with the board. Québec remained poorly represented. Another weakness of the board at this time was that an important new component of the heritage movement concerned with the preservation of historic architecture was not represented. This really was a missed opportunity because the parks branch had access to depression relief funds which enabled it to embark on projects designed to provide employment in the provinces. Meanwhile, the provinces were taking initiatives in the area of architectual preservation. In Nova Scotia, Québec, Ontario, and British Columbia the provincial government contributed to the recognition and conservation of historic buildings. As before, both the branch and the board were criticized by a significant part of the heritage community, while the branch lost a valuable source of historical advice.

The new decade brought an economic crisis, and a new Conservative government intent on reducing public spending. Accordingly, the operating budget of most government agencies, including that of the parks branch, was severely cut. As a result, in 1931 the position of the chief of the historic sites division was abolished, and Pinard was given early retirement.[1] The historic sites division disappeared, and Pinard's former assistant, G.W. Bryan, was left to carry on alone. This also left a relatively minor civil servant to represent the program in the department. This had the immediate effect of crippling the capabilities of the agency. Without divisional status it had a greatly reduced operating budget and became more than ever dependent on the services of other divisions and government departments to carry out its work. Bryan, it seems, lacked Pinard's competence and he does not appear to have been as effective as his former chief in seeing to the needs of the board which he now served as secretary. Certainly Howay was disparaging about his performance, commenting to Cruikshank that "a more useless person as a secretary it would be hard to find; or one more lacking in initiative or in the ordinary equipment of a secretary."[2] The most significant impact of government austerity for the board was the reduction of the number of monuments that could be erected by the department, thereby severely limiting its commemorative program. Whereas the branch had endeavoured to erect five plaques annually in each region, during the early 1930s it sometimes had difficulty erecting one.[3] This increased the backlog of sites so that of 292 designations made by the board by the end of the fiscal year 1933–4, only 200 had been commemorated.[4] As many of the commemorated sites reflected the regional inequities which had existed before 1923, this delay was especially noticeable outside of central Canada.

The early 1930s formed a critical juncture for the board. New challenges were being raised by the heritage movement, and new opportunities presented by the depression relief work of the branch. Unfortunately, its position at this time was uncertain. The National Parks Act of 1930, which accorded formal status to historic sites and parks, had failed to give the board the legislative footing needed to operate as an independent agency. Its position was further eroded with the departure of Pinard.

Internally, the board showed signs that it lacked the will to rise to new challenges. Conditioned by the frustrations of the 1920s, and the limited personal agendas of some of its leading members, the board was dominated by a mood of inertia. By the early 1930s members were questioning the future of its commemorative programe. As early as 1928 Webster had suggested to Harkin that the list of potential national historic sites in New Brunswick had been largely completed.[5] Again, in 1932 the branch acknowledged Webster's recommendation that "all historic sites of major importance have now been dealt with."[6] About the same time Crowe made a similar assessment of sites in his province. In 1931 he noted that "apart from the Louisbourg Park, where it is my hope that reasonable expenditures will continue from year to year, I should suppose the expenditures in the Province should gradually diminish."[7] That year both Crowe and Coyne resigned from the board, no doubt feeling that their major objectives had been achieved. By 1935 Howay, too, believed that the commemorative program had largely carried out its task of marking all the important sites in the country. In a letter to Cruikshank he said, "[t]he fact is that we have pretty well covered the field in every province. Shall we go on? and put the relatively unimportant on the same level as the important? or shall we just quit?"[8] That Howay also contemplated resigning the next year demonstrates the mood of the board in this period. Its immediate problem lay with justifying its commemorative work. It had little inclination to expand its activities into the area of preservation.

Fortunately for the board, the appointment of new members provided new energy and ideas. In 1931 D.C. Harvey (1886–1966) replaced Judge Crowe as the Nova Scotia representative. At forty-five, Harvey was the youngest member ever appointed and represented a new generation of professional historians. Born in Prince Edward Island, he attended Dalhousie University and then went to Oxford as a Rhodes Scholar. Subsequently he taught history at colleges in Winnipeg and Vancouver before moving to Halifax to take over the newly re-organized provincial archives in 1931. There he served as a part-time professor at Dalhousie and became active in local historical work. Harvey had well-defined views about a distinctive Nova Scotian historical perspective. In an essay "History and its Uses in Pre-Confederation Nova Scotia," he wrote that history confirms "the conviction that Nova Scotians are the equals of any other British subjects and the aim to have them incorporated in the British system rather than the Canadian or American."[9]

Harvey's argument for a distinctive regional history complemented Webster's own sense of the centrality of Maritime history, and the two collaborated on a number of commemorations. Harvey agreed, for example, that the founding of Saint John and the province of New Brunswick were important events in Canada's constitutional history.[10] But Harvey's imperialist views also found favour with Cruikshank, and the General strove to defend his younger colleague from some of Webster's more tyrannical gestures. Thus Harvey was encouraged to represent the views of Prince Edward Island which until then had been the domain of Webster. Aided by his own knowledge of that province's history and spurred by local concerns for better representation in the pantheon of national historic sites, Harvey sponsored the designation of three new sites in the province: the Roma Settlement, the first organized land survey, and Jacques Cartier. Harvey also differed from Webster in favouring the preservation of the Halifax Citadel.[11]

Also in 1931, James Coyne retired and recommended the appointment of Fred Landon whom he knew through local heritage circles.[12] Landon (1880–1969), like Harvey, differed from his predecessor in being more in the mold of the new generation of historians. He had an MA in history from the University of Western Ontario and taught Canadian and American history there as well as being the university's chief librarian. At the time of his appointment to the board Landon's reputation was largely based on the five volume history of the province of Ontario which he co-authored with Jesse E. Middleton.[13] Subsequently he contributed an important volume to the series of publications on the relations of Canada and the United States, *Western Ontario and the American Frontier* (1941). He was successively president of the London and Middlesex Historical Society (1918–20), the Ontario Historical Society (1926-8), and the Canadian Historical Association (1941–2). Like most of his colleagues on the board, he became a member of the Royal Society of Canada.[14]

Landon's view of Canadian nationhood differed from that of Coyne and Cruikshank in that it looked to the positive aspects of American influence on Canadian development. In his conclusion to *Western Ontario and the American Frontier* he noted that the main influences on the development of the region's society were geography and immigration. Wars and battles had little importance in his view of historical development. He saw Ontario society as an ethnically diverse community that allowed the formation of institutions based on the best of the old and the new worlds. Moreover, he argued that "the question of loyalty itself has no place in this discussion and it is probably true that descendants of Loyalists have emphasized much more than did their forefathers the extent of the sacrifice made on behalf of British allegiance."[15] Landon also differed from Cruikshank and Coyne in his recognition of the progressive nature of the rebellions of 1837–8. Where the other Ontario representatives viewed the incidents of 1837–8 as another example of loyal men defending the empire against alien ideas,

Landon saw the episode as a step toward responsible government. Shortly after his appointment to the board, therefore, he recommended commemorating the site of the Battle of Windsor, an episode in the 1837–8 rebellions. But although Howay agreed that it was an important event on the road to political liberty, the rest of the board voted against it.[16] Further nominations of this kind did not come until after the departure of the General.

Before the war the heritage movement had been a fairly cohesive unit, finding expression in the Royal Society of Canada and the Historic Landmarks Association. Organizations at both the local and national levels were dominated by amateur historians, people of educated backgrounds but with little or no specialized training in the field of history. These people tended to regard the past as an integral part of the present, co-existing with an awareness of literature, art, and general democratic principles. History, then, was the preserve of a cultural élite which sought to recreate the past according to ideals of their *weltanschauung*. In the prewar heritage movement historians could comingle with poets, politicians, prelates, and other members of this élite in an atmosphere of common understanding.

Although this interdisciplinary élite continued to be an important factor in the heritage movement after the war, just as it continued to dominate the Royal Society, a new generation of professional historians was emerging that, by the 1930s, was exhibiting a distinct orientation toward specialized themes. Members of this new generation had postgraduate degrees, increasingly with a PHD from an American or Canadian university and engaged in research that was narrowly focused and writing that was often interesting only to a small scholarly audience. Whereas in the 1920s academic historians were typically men like George Wrong with non-specialized degrees, history departments in the 1930s became dominated by men such as Chester New at McMaster, R.G. Trotter at Queen's, and W.N. Sage at the University of British Columbia. Generally the two generations co-existed amicably in organizations such as the Royal Society and the Canadian Historical Association, although there was some tension as the younger experts flaunted their newly acquired professional pretensions, and the old guard wondered if some of them were indeed gentlemen.[17] Also, the new group was even less inclined than the original to be interested in preservation. The concerns of this new generation tended toward abstract principles of national development. Conditioned by the nature of university teaching, academic historians had little interest in material culture.[18]

The chief result of this generational distinction was a diffusion of the aims of organizations like the Royal Society and the Canadian Historical Association, the traditional vehicles of the heritage movement. They no longer undertook to be largely concerned with public aims such as heritage preservation but, as organs of the new professionalism, sought out increasingly specialized audiences. Throughout the 1930s, although the *Canadian Historial Review* and the annual reports of the Canadian Historical Association continued to provide accounts of the activities of the parks branch in identifying and preserving national historic sites, they reported little other preservation news, and the articles were largely written by professionals.

Yet despite the potential for acrimony, because of this gap between the old guard and the new, relations between Cruikshank and Landon on the board were largely harmonious. While Cruikshank was chairman Landon did not challenge his proposals and instead supported his drive for increased loyalist commemorations in 1935. Such solidarity should not appear overly surprising given the smallness of the Canadian historical community at this time. As practising historians, both amateurs, like Webster and Cruikshank, and professionals like Harvey and Landon, met as equals at meetings of the Canadian Historical Association and the Royal Society as well as on the board. While they may have been opposed ideologically, both Cruikshank and Landon were part of a tightly knit group of Ontario heritage activists joined in a common cause and so had little choice but to get along. Thus even in his revisionist history of western Ontario, Landon graciously acknowledged the advice of the late General.[19] Both Harvey and Landon, then, while representing the new generation of professional historian, still had strong ties to the old guard. Nonetheless, their membership encouraged the board to broaden its views beyond the personal agendas of the older members.

The appointment of Edouard-Fabre Surveyer (1875–1957) in 1933 to represent Québec posed some difficulties. Unlike the two previous appointments, Surveyer was of the traditional antiquarian school. But as he remained on the board until 1956, he provided more consistent representation for French Canada than any previous Québec representative since Benjamin Sulte. Born in Montréal, he was educated at McGill and Laval universities before being called to the bar. He had a distinguished career as a jurist and was professor of criminal law and procedure at McGill before his appointment to the Québec Superior Court in 1920. Like many francophone professionals, Judge Surveyer took a great interest in the history of his province. He was a member of a number of historical societies and published articles on local history as well as the prosaic but respected book, *The First Parliamentary Elections in Lower Canada*. As a member of the board, Surveyer considerably bolstered the representation of Québec. He was a conduit to the board of the views of local heritage groups and he maintained reasonably friendly relations with his anglophone colleagues, particularly Webster.

But Surveyer's credentials were not entirely what they seemed and he probably did not have the complete confidence of Québec nationalists. He had been unable to get himself elected to section I (that is, the French-speaking section of the Royal Society), ostensibly for his pro-English views, so in 1930 he sought the help of Webster to get elected to section II (the English-speaking section).[20] Webster apparently achieved this unusual selection, which may have given the judge a certain cachet among anglophone Québecers, but would not have helped his credibility in French Canada. It was also Webster, it seems, not Québec heritage groups, who secured his appointment to the board.[21] These connections did not make Judge Surveyer an ideal representative of the board in French Canada. His influence on the board was slow to be felt, partly because of his diffidence in nominating sites. But later it would become apparent that he was strongly disposed to champion the perspective of French Canada in the commemoration of a national history. Like Harvey and Landon, then, he harboured a potential for wider designation of national historic sites, while maintaining overt sympathy with the traditional views of the board.

Reinforced in this way, and with the mixed assets of the other members, the board met the challenge of the 1930s obliquely. The backgrounds of the new members did not dispose them to raising the profile of preservation on the board. While representing the hitherto largely ignored concerns of professional historians and French Canada, they were too strongly tied to the older members to challenge the direction of the commemorative program. Thus commemorations in the early 1930s followed the themes established in the 1920s. In the Maritimes, forts and "first things", in Québec more forts and early communications, and in Ontario more loyalist designations defined the activities of the board. In the west Howay strove to give better representation to sites east of British Columbia, but many of these reflected traditional themes of exploration and the North-West Rebellion of 1885. More than intellectual limitations kept commemoration in this rut. Abstract aspects of Canadian history were difficult to realize in an historic sites program and financial restrictions of the first years of the depression meant that new commemorations became even more difficult to erect than in the past. So, even with new historical themes being overshadowed by the weight of past concerns, they had little chance of being reflected in a program that was barely functioning. Still, efforts were made in the early years of the new decade to broaden the scope of the board.

As before, Harkin tried to shift the concerns of the board away from hackneyed themes to consider other subjects worthy of national commemoration. He proposed to the members "that action be taken to have memorials erected to commemorate the constitutional changes in the progress of the Dominion."[22] Howay was in favour of this proposal and undertook to investigate how it could be implemented. But the vagueness of the proposal made appropriate sites difficult to find. In 1934 Howay proposed that constitutional development be commemorated in a long but succinct inscription noting the principal stages in the evolution of the Canadian nation, and that it appear on plaques erected on the Parliament buildings in Ottawa and in the provincial capitals.[23] But this proposal was not supported by the rest of the board which had already dropped constitutional changes as an impractical subject to commemorate.[24]

Another avenue the board explored to enlarge the scope of its commemorative program was the creation of a secondary marker. Until the 1930s all national historic sites were commemorated by a standard bronze plaque bearing an inscription explaining its importance. While the criteria for the designation of sites continued to be vague, the board remained preoccupied with the notion of *national* significance, and many board members – particularly Cruikshank, Howay and Webster – felt that most of the obvious sites had already been identified. Unless the criteria could be broadened, something these members did not wish to do, a solution to the problem of widening the commemorative program was to create a lesser category of national historic site to be marked by a smaller bronze plaque than the standard version. During the days when the branch had difficulty in maintaining its normal quota of commemorations, these cheaper monuments, designed to be attached directly to a building rather than the usual stone cairn, became additionally attractive. The idea for a secondary tablet had been mooted back in 1925 as a means of commemorating sites of lesser importance, and Howay saw it as an appropriate means of commemorating prominent Canadians.[25] Although this proposal was not acted upon by the board, Webster inveigled some secondary plaques for his Fort Beauséjour development in 1929.[26]

These, however, did not mark additional historic sites, merely additional themes within the historical park.

The idea of secondary commemorations was raised again in 1935 by Judge Howay in a letter to Cruikshank. Howay believed that the main obstacle preventing the establishment of a lesser category of historic sites was Harkin, for he sensed that the commissioner saw lesser markers as cheapening the primary designations. He disagreed with this view.

> In fact most of the things that are left are pretty small potatoes; and it seems to me quite out of keeping to give them standard tablets; but what are we to do if Mr. H. will not? We shall stultify ourselves by giving undue importance on standard tablets to what is secondary. Some of the things you are offering in connection with the War of 1812 (and some that are already done) should be secondary, if noted at all. And that is true all round. The fact is that we have pretty well covered the field in every province. Shall we go on? and put the relatively unimportant on the same level as the important? or shall we just quit? If there is no secondary tablet then in my view our work is about done.[27]

The board had taken a step toward introducing secondary commemorations in 1934 when it approved a list of fifty prominent historical figures.[28] But by 1936 the branch had still not implemented a program of secondary plaques. While Harkin's lack of enthusiasm for these lesser monuments may have been a factor in the inactivity of the branch, the miniscule budget allotted commemorations inhibited new initiatives in this field.

By the 1930s a new force had emerged in the heritage movement concerned with preserving historic architecture. The shift of the membership of the Royal Society and the Canadian Historical Association away from heritage concerns opened the field to other specialized groups concerned with promoting aspects of heritage conservation. Chief among these were architectural associations which advocated a greater awareness of historic architecture. Even before the First World War the Association des architectes de la province de Québec had tried to encourage interest in the province's old buildings.[29] Following the war there was increasing recognition of traditional Québec architecture as part of a collective heritage. This awareness was promoted by the faculty of the McGill University School of Architecture during the 1920s and 1930s. There a group of British-trained architects, principally Percy Nobbs and Ramsay Traquair, sought to incorporate traditional designs into modern buildings. While in Britain and much of the rest of Canada this same impulse resulted in houses designed to resemble Cotswold cottages or Tudor mansions, it inspired the McGill faculty, especially Traquair, to study vernacular Québec architecture as a particular style. During the 1920s students in the faculty of architecture were required to produce measured drawings of old Quebec buildings, and the school accumulated a large inventory of plans and elevations of public buildings, churches, and residences, many from the French régime. Traquair used this collection to supplement his own considerable body of knowledge in writing *The Old Architecture of Quebec* (1947). He approached his subject, not as an antiquarian, but as an architect, and its significance was valued for its present appearance, not for its past associations.[30] Well before his book appeared, however, the work of the McGill faculty had been recognized for its identification of a distinctive French-Canadian culture. In praising the work of the McGill faculty in 1926, the Montréal amateur historian and future member of the board, Surveyer, said: "[t]hey have realized, to an extent that cannot be appreciated without careful study ... what glory there was in the architecture founded by the ancestors of Mr. Montpetit and by mine – the first settlers of this country."[31]

In Ontario there was a similar interest in the identification of what was considered to be a distinctive provincial style. Here the University of Toronto school of architecture played a role similar to its Québec counterpart in promoting awareness of heritage buildings. As at McGill, one of the leading exponents of a distinct provincial architecture was Eric Arthur, a British-trained architect interested in incorporating traditional architectural styles into modern designs. Arthur was born in New Zealand and then trained at the University of Liverpool before joining the University of Toronto faculty. During the 1920s he published a series of articles in the journal of the Royal Architectural Institute of Canada under the general heading "The Early Architecture of the Province of Ontario." His research, like Traquair's, also supported by measured drawings prepared by his students, culminated in his *Early Buildings of Ontario* (1938). Where Traquair had identified a particular building type associated with the Quebec habitant, Arthur identified a distinctive loyalist style of dwelling. As in Québec, this revelation fitted in well with the regional version of nationalist ideology.

An awareness of a distinctly provincial architectural heritage led to measures being taken for its preservation. Not surprisingly, Québec was the first province to enact legislation for its protection with the passage in 1922 of "An act respecting the preservation of monuments and objects of art having an historic or artistic interest," (12 Geo. v, c.3). The Commission des monuments historiques de la province de Québec was established to implement this legislation and it compiled inventories of heritage resources such as historic and devotional monuments, old churches, French forts, and old houses. Although this commission functioned only for a short time, producing three progressively diminutive annual reports, its work was continued by Pierre-Georges Roy, the provincial archivist and secretary to the commission. In 1927 he compiled *Vieux manoirs, veilles maisons*, a collection of photographs of traditional Québec residences, mostly from the French régime. The first Québec program was mainly an inventory, and buildings on its list were not protected by any legislated sanction. Ontario did not enact heritage legislation until after the Second World War, although groups such as the Architectural Conservancy of Ontario, founded in 1933, sought to preserve loyalist houses, especially in Niagara-on-the-Lake. The only other province to enact heritage legislation in the interwar period was British Columbia which passed an act in 1925 aimed at preserving Indian artifacts. This act, however, like the Québec legislation, was ineffective, and without money to acquire historic property, little could be done to prevent its loss. Nonetheless, the principle had been established by the 1930s of provincial governments protecting historic property. This further served to diffuse the heritage movement which until then had focused almost entirely on the federal government.

In Ontario the provincial government became a leading participant in the heritage field following a series of initiatives by the Niagara Falls Parks Commission. Although its larger projects of restoration and reconstruction were not undertaken until later in the 1930s, its involvement reflected an earlier trend of provincial interest in heritage preservation. Originally formed in 1895 to oversee the development of public land surrounding Niagara Falls, the Queen Victoria Niagara Falls Parks Commission came to assume control of a number of commemorative monuments including the one to Brock at Queenston Heights and memorials to Laura Secord and the Battle of Lundy's Lane. Subsequently, the commission incorporated historic properties into its sprawling park system.

During the 1930s the commission sponsored four important projects of restoration and reconstruction. The most ambitious of these involved the rebuilding of Fort George at Niagara-on-the-Lake whose fourteen buildings had been destroyed during the War of 1812. An historian was hired to locate documents related to the building and interpretation of this site, and between 1937 and 1940 the fort was reconstructed to appear as it did between 1796 and 1799. This involved reproducing bastions, stockades, and eleven of fourteen buildings according to original specifications and furnishing a number of rooms with period reproductions and historic artifacts to demonstrate the function of the buildings.[32] Another monument from the British military establishment at Niagara, Navy Hall, received unique treatment. Originally a large wooden building, it survived in a ruined state, having been moved from its original site. The commission had it reinstated to its former position, then erected a large stone building in a suitable antique mode to protect the ruins from further deterioraton. A third project was the restoration of an early nineteenth-century stone mansion at Queenston once inhabited by William Lyon Mackenzie. The fourth undertaking was the partial reconstruction of Fort Erie, also destroyed during the War of 1812.

The development of these historic properties not only fitted into the goal of the Niagara Parks Commission to beautify the area by contributing an antique or romantic flavour to the acres of lawn and scenic drives, but was justified by encouraging another goal of the commission, tourism.[33] Although the commission focused on developing Niagara Falls, one of the world's great tourist attractions, it undoubtedly believed that a series of supporting attractions would be of benefit. The examples of Fort Ticonderoga and Williamsburg as meccas for the new era of motorized tourists further encouraged a commitment to bringing to life buildings of the past.

The success of the Niagara Parks Commission's heritage properties was not lost on the Ontario Department of Highways which was also intent on developing tourism in the province. In 1936 the department initiated the restoration of Fort Henry in Kingston. This huge structure, begun in 1829 and completed in 1842, was an important link in the British defensive network in North America and was known as the Citadel of Upper Canada. Although it saw some use during the rebellion of 1837, the fort was obsolete soon after it was built, and following the withdrawal of the British garrison it became derelict. Well built of dressed stone, it remained largely intact as a well-known local landmark. All that was needed was some landscaping and repair to its masonry. Such a labour-intensive project appealed to the federal government which, independent of its national parks branch, was prepared to support the endeavour. The work was probably directed by people experienced in the Niagara Parks Commission historic projects. Ronald Way, the historian responsible for much of the research on Fort George, was also employed to find the original plans of Fort Henry. He subsequently commented that "[c]areful research and painstaking workmanship were combined to produce results which professional historians concede to be as accurate as any similar achievement in North America."[34] The restorations went beyond repairing walls and roofs; interiors were refurbished and furnished with period reproductions and historic artifacts. During the summer months the fort was animated by students dressed in the uniform of British imperial troops of the period and called the Fort Henry Guard. The fort was opened in 1938.

These developments demonstrate that the heritage movement in Canada was significantly different in the 1930s from what it had been before the war. Preservation and heritage development had a greater public profile than commemoration which was no longer viewed with quite the interest that it once was. Unfortunately for the federal heritage program, this new voice was not represented on the board. This lack of representation of the new heritage interests is reflected in the unsympathetic response of the board to their proposals.

While the board was prepared to act on a number of requests from local societies and individuals for the commemoration of historic sites it consistently rejected proposals to become involved in the preservation of buildings. In 1933 it reiterated its policy of not commemorating old houses for their architectural merit.[35] And, while it was willing to recommend the erection of a plaque to commemorate the former home of Alexander Graham Bell in Brantford, Ontario, the board ignored the request from a citizen's committee for the government to provide an annual subsidy for the upkeep of the property.[36] Regularly through the 1930s both the board and the branch turned deaf ears to a number of similar requests.

While this policy reflected the board's awareness of the limited funds available for preservation and the fear that acceding to one worthy request would invite an avalanche of others, it was also indicative of the continued lack of concern on the part of most members for preservation. Following the opening in 1933 of colonial Williamsburg, the restored Virginia townsite funded by the Carnegie endowment, Harkin was intrigued enough to canvas the board members to see if they favoured the undertaking of similar projects in Canada. Harvey's comments were probably a relatively mild response. "In reply ... I may say that whilst I was very much interested in the account of what was being done there, I cannot but feel that it is a case of misdirected energy and extravagance in the name of historical romanticism. It seems to me that we can spend our money more wisely in the preservation of actual historic sites, and in making museums from historical spots than by tearing up modern streets and houses in order to reproduce the atmosphere of bygone days."[37] Cruikshank did not even favour the establishment of museums. Arguing against the restoration of Fort Malden in 1935, he said: "[a]s you are aware, my opinion is that the creation of small local museums at more or less out-of-the-way points, which have certain historic interest, is a mistake, and to a certain extent a waste of public money."[38] Webster, on the other hand, while being generally in favour of restoration, museum,s and historical reconstruction, showed little interest in projects beyond his domain.

The members of the board, then, had some well-defined and reasoned opinions on heritage development and preservation. But because the board as a whole remained aloof from this side of the program, its views had little effect on overall policy. Instead, the branch was left to develop policy on its own, and during this period its own organizational weakness mitigated the possibilities of formulating a clear and consistent policy. This was unfortunate as such a policy was necessary to meet the new opportunities for the preservation side of the historic sites program presented by the 1930s.

CHAPTER FIVE

New Deals, 1932–40

By the 1930s the federal heritage program seemed destined for euthanasia. It had been thwarted in many of its initiatives to develop a comprehensive policy for preservation and commemoration, its advisory board looked tired and unimaginative, and the parks branch appeared to have forgotten it. But just when it seemed that the program was about to pass on, it became rejuvenated by two extraordinary events of the decade: the depression and the ensuing government relief projects, and a bureaucratic reorganization that placed the whole national parks agency into a fresh new departmental structure. These developments placed the heritage program more firmly under the control of the branch, while further weakening the power of the advisory board which remained preoccupied with commemorations.

DEPRESSION RELIEF PROJECTS, 1932–6

The depression brought opportunities for preservation projects because while the government initially reduced departmental expenditures, it soon compensated by introducing relief measures in the form of public works spending. The nature of the operations of the parks branch allowed it to become a major participant in these projects. In his examination of federal government welfare policy during the depression, James Struthers has described the central role played by Andrew McNaughton's relief camps, organized under the auspices of the Department of National Defence in 1932. The objective of the camps was to provide unskilled work in areas of federal responsibility. Work on or around military fortifications was therefore an obvious project, and McNaughton proposed restoration work on the citadels at Halifax and Québec as well as clearing along the route of the proposed Trans-Canada airway.[1] Camps were formed, initially limited to two thousand men, to carry out these projects by the end of the year. Yet it is not generally known that the parks branch also embarked on projects as part of the same policy to give employment to able-bodied men. During the fiscal year 1931–2 the branch received special appropriations of $875,000 for relief projects in national parks.[2] Most of this activity involved road building in the Rocky Mountain parks, but as McNaughton came to regard the restoration of historic fortresses as suitable projects, so too did Harkin. In this new climate preservation projects previously disregarded by ministers and senior department officials as being too costly were now being reconsidered in another light. Here at last was the opportunity to effect a meaningful policy of preservation. And here, too, was a chance for the board to broaden the scope of its activity.

Lacking a clear policy on preservation, but with money to spend, the branch was susceptible to the well-defined objectives of local enthusiasts. During the 1920s the role of the branch had been largely one of merely restraining these enthusiasms in light of budgetary limitations. The scope of possible preservation was therefore narrowly defined. Although the branch was guided by the views of McLennan, Fortier, and Webster, it could only undertake minimal development. In 1931 it seemed as if even this small capability would be curtailed, lessening the scale of already modest plans. But public works relief money being made available after 1932 gradually began to affect the capacity of the branch to undertake more costly heritage projects. A difficulty lay in the lack of direction and expertise at the centre. The ideas for restoration largely rested with peripheral groups. The board, which should have been at the centre of this activity, chose not to become involved. Too large to be undertaken by Bryan's small historic sites section, preservation work became the concern of the engineering division and other government departments. Thus, although greatly expanded, preservation work in the 1930s was characterized by a lack of focus and clear policy.

At Louisbourg, for example, the engineer seconded from the engineering division came under the watchful surveillance of the local committee. Crowe remarked to Webster that the engineer had made gun carriages of stone to hold two cannons situated in front of the temporary museum. "The whole thing is incongruous, and is remarked on by visitors who have knowledge of such things. A naval gun carriage should have been made. I remonstrated with

Roberts, the Engineer then in charge of the works, but without results. He said he knew nothing about naval carriages. A half hour's search in the Parliamentary Library or at the Defence Department would have procured the exact specifications."[3] With little expertise emanating from Ottawa, the Louisbourg local committee undertook to keep development historically authentic. They also undertook to promote the site as a tourist attraction, and during 1933 Crowe and McLennan were busy planning the publication of a guide book.[4] The centrepiece of the Louisbourg development in this period, however, was the museum, and responsibility for its construction was largely removed from the branch.

The Louisbourg committee had long complained about the temporary museum which they said was a firetrap, but the branch had been unable to obtain the authority to construct a new building. Then in 1932 Webster thought that he could obtain funds from the Carnegie foundation for museum buildings at Louisbourg and Fort Beauséjour and subsequently urged the branch to submit proposals. Although nothing came of the Carnegie proposal, the passage of the Public Works Construction Act (24–25 Geo. v, c.59), and the commitment of Bennett's government to relief projects made the construction of these buildings a viable project. Plans were prepared by W.D. Cromarty of the architectural division, and construction carried out under the supervision of the Department of Public Works during the summer of 1935. The same design was used for the museum building at Fort Beauséjour, which was put up at the same time. The Louisbourg museum came under the personal control of the McLennan family. Katherine McLennan, the senator's daughter, was appointed honorary curator, and many pieces on display were from the McLennans' own collection. Similarly Webster took charge of the Fort Beauséjour museum, personally selecting and acquiring artifacts for display there.

Elsewhere in the Maritimes the Public Works Construction Act enabled the reconstruction of the officers' quarters at Fort Anne for the museum organized by the superintendent, L.M. Fortier.[5] In Halifax the Department of National Defence undertook the restoration of the Prince of Wales's Martello Tower, acquired as a national historic site in the 1920s.[6] The campaign to have the government construct a replica of the Port Royal Habitation was resurrected in 1934 by Mrs Richardson and Dr Webster. This time the minister seemed sympathetic to the proposal. He recommended setting aside funds for the project and in June 1935 asked the parks branch to prepare detailed plans and estimates. The board was deliberately excluded from this venture, and Harkin advised his superiors that "[a]ny action on the lines suggested by Dr. Webster I think would have to be taken by the Department independent and distinct from the Historic Sites and Monuments Board because the Board has already marked this site and does not commit itself to reconstruct."[7] It was Webster who asked to make the presentation to cabinet and in August the project had received widespread support. But there were other difficulties, besides the diffidence of the board, which precluded government action. Most of the necessary real estate was owned by an unsympathetic farmer who wanted to sell rather than donate the property, and at this point the government seemed reluctant to spend money on real estate. Then, with the calling of the October 1935 general election, the whole project was allowed to drop.

In Québec branch spending on heritage development was largely determined by the enthusiastic curator of Fort Chambly. He and his predecessors had worked for a number of years building up the site, not as an example of military life, but as a centre of traditional Québecois culture. The branch had not interfered with this line of intepretation and in 1928 had agreed to build a museum within the fort's walls to help develop this function. Lack of funds postponed the implementation of this scheme until 1934 when plans were prepared by the Department of Public Works and the building was erected using funds from the Public Works Construction Act. To help the curator better develop his collection of French-Canadian folk art, the parks branch obtained the secondment of Marius Barbeau, a leading expert on the subject, from the national museum in Ottawa.[8]

In Manitoba the initiative came from the provincial government which urged the restoration of Prince of Wales's Fort. In 1930, with the near completion of the Hudson Bay Railway and the townsite of Churchill, the Manitoba government sought to develop tourism in the area. As the fort was a major potential attraction, the government applied pressure on Ottawa to develop the site for visitors. Consequently, in the fall of 1931 the minister of the Interior, Thomas G. Murphy, asked Harkin to prepare plans and estimates for the rehabilitation of the ruined fort. These were subsequently provided by the engineer of the Department of Railways and Canals resident at Churchill, but the estimated cost of twenty thousand dollars discouraged immediate action.[9] Continuing political pressure and money made available through the Public Works Construction Act combined to make the plan feasible, and between 1935 and 1937 rudimentary restoration was carried out to turn the ruin into a tourist attraction. The walls were built up and given a concrete cap, the gateway was refashioned using cement scored and coloured to resemble ashlar, and the spiked guns were salvaged and placed on new carriages. This work was carried out under the supervision of the resident engineer of the Department of Railways and Canals who established a camp of seventeen men plus a cookhouse for the summer season to expedite the work.[10] His results met with widespread approval.

In British Columbia at Fort Langley, the old Hudson's Bay building, was the only other structure preserved by the branch west of Manitoba. It was being operated as a museum by the Native Sons of British Columbia. Urgently needed repairs were carried out using funds from the Public Works Construction Act, and a supplementary act passed in 1935. The wall was rebuilt, "the new timbers being framed and scored with a broad axe to resemble the original as much as possible."[11] Another wall was extensively reconstructed using salvaged material, the walls were reinforced with iron tie rods and the whole building placed on concrete piers. The building remained in this condition until 1958 when it was incorporated into a large reconstruction project launched to coincide with the province's centennial celebrations.

These developments attracted favourable publicity to the federal historic sites agency. Restoration projects at Louisbourg, Beauséjour, Fort Anne, Fort Chambly, Churchill, and Fort Langley raised the profile of historic sites within the branch as well. After years of struggle to win government commitment to preservation, the coffers had at last been opened. The Historic Sites and Monuments Board, too, had realized long-held goals in this period as many of its members acknowledged that much of the original list had been commemorated. It had some success in broadening its criteria and managed to agree on a secondary program of commemorations. Ironically, increased activity in this period was marked by a corresponding decline in the influence of the board and the branch. The board, especially, missed an important opportunity to influence preservation policy. These tendencies continued in the ensuing years. Although the board would still be identified with the whole historic sites program, it would confine itself to designating and commemorating historic sites. Preservation rested with the parks bureau which underwent a major reorganization in 1936.

A NEW DEPARTMENT, 1936–40

During the early 1930s the parks branch had fared only moderately well under the administration of the Bennett government. Its home, the Department of the Interior, had lost considerable power and prestige following the transfer of natural resources to the western provinces, and with the deepening of the depression the work of the parks branch lost some of its relevance. This loss of prestige on the part of the branch was aggravated by a waning of Harkin's personal influence with the government. Recruited into the civil service by Clifford Sifton, he was a known Liberal placeman, and although he had established a solid reputation as a conservation expert and had worked well with Arthur Meighen during the Borden régime, he was regarded warily by the new Conservative government of 1930 which, after years in the political wilderness, suspected Liberal influence behind every bureaucrat.[12] Consequently, despite Harkin's record in developing the parks branch into an important conservation agency, he was passed over for promotion by the Bennett government and never rose to assistant deputy minister. In such a milieu Harkin would have had even less influence with the minister of the Interior, the Honorable Thos G. Murphy, than during the previous régime.

The prestige of the parks service and its commissioner seemed to plummet after the election of 1935. With bureaucratic efficiency as one of its central concerns, the new King administration combined four departments – Interior, Mines, Immigration and Colonization, and Indian Affairs – into a single Department of Mines and Resources with Thomas A. Crerar as its minister. The new department was organized into five main branches: mines and geology; lands, parks, and forestry; surveys and engineering; and Indian Affairs and immigration. Within this scheme parks assumed an even lower profile than in the Department of the Interior. It was buried in a polyglot branch with three disparate bureaus: Northwest Territories and Yukon affairs, land registry, and the Dominion Forest Service (see appendix 1). It had not only lost its branch status, but its dynamic engineering division had been taken away to form the basis of a separate surveys and engineering branch. The former chief engineer of the division now outranked the head of the national parks bureau. With such a large reorganization there were too many senior officials vying for too few positions. The former assistant deputy minister of the Interior, R.A. Gibson, was made director of the lands, parks, and forestry branch. Harkin was offered the position of controller of the national parks bureau, which would have meant a reduction in rank.[13] Instead he chose early retirement and left the public service in 1936 at the age of sixty-one. So ended an era in the history of Canada's national parks, the conservation movement, and the historic sites program.

But the situation was not as critical as it initially seemed for either the parks service or its historic sites agency and it actually began to improve after 1936. For one thing the new actors on the stage expressed a welcome interest in historic sites. Crerar seems to have been personally interested in the subject. He visited the important heritage projects in the east and was the first minister to attend a meeting of the Historic Sites and Monuments Board.[14] Unlike ministers of the Interior, Crerar regarded the whole of Canada, not just the west, as his domain, and under his administration the parks bureau began to pay attention to the east in a way that it had not done before. As historic sites were considered to be an important eastern resource, this shift in emphasis would benefit the historic sites program.

The program also benefited from an internal reorganization following the move of the parks service to the Department of Mines and Resources. During the early 1930s the historic sites program had suffered from the lack of a strong administration. But in 1936 Williamson assigned the program to W.D. Cromarty, the bureau's chief architect. Williamson had been with the parks branch since before the war and as Harkin's principal lieutenant had a detailed knowledge of the various facets of the parks mandate, including historic sites. He continued to remain involved in major decisions and took Harkin's place on the board, but increasingly it was Cromarty who took charge of historic sites matters.[15] This responsibility increased following Williamson's death in 1941. His successor, James Smart, chose not to sit on the board, and Cromarty as secretary formed the sole liaison between the board and the department.

After a long association with the national parks service as an architect and planner, Cromarty had some well-formed views about historic site development. Trained in England, he had immigrated to Canada in 1912 where, after a stint in private practice in British Columbia, he joined the department of architecture at the University of Alberta and was for a short spell a colleague of Ramsay Traquair prior to the latter's move to McGill. He joined the parks branch in 1921 being assigned to the town planning division where he became an associate of Thomas Adams.[16] The town planning division had been instrumental in establishing a distinctive park architecture, one that was intended to convey a federal message while enhancing rather than distracting from the natural scenery. Harkin once wrote that "I feel that everything our engineers construct in the Parks should be dominated by the spirit of beauty,"[17] and it was the job of Cromarty to see that this objective was achieved. An example of his work was the upper hotsprings bath house designed for Banff in 1932. With roughly hewn stone, a steeply pitched roof, and half-timbered dormer, the building achieved a certain rustic elegance not unlike an English country house from the same period. A parking lot was artfully concealed so that the facilities were convenient for the motorist yet the grounds were not disfigured by their modern conveyances.[18]

This approach to parks development helped determine Cromarty's perspective on historic sites improvement. He believed, like Thomas Adams, that proper landscaping was an important aspect of heritage development, and that preservation meant encapsulating historic ruins in a park setting. He also believed that improvements should reflect the antique flavour of the site while not necessarily reproducing original buildings. It was Cromarty, for example, who had designed the museums at Louisbourg and Beauséjour which were vaguely described as being in the French Château style. With Cromarty in charge, the visual or aesthetic approach to heritage development had a strong advocate.

The interest of the minister, the concern of Williamson, and the experience of Cromarty were all important factors in shaping a federal heritage policy. But a policy was nothing without money, and the historic sites program had been crippled by a lack of financial commitment. This situation changed with the advent of Crerar who soon emerged in King's cabinet as an advocate of larger departmental expenditures.[19] As a result, the parks budget for the fiscal year 1936–7 actually increased over that of the previous year when it had branch status. With the exception of the war years it increased steadily after that, and whereas spending for 1936 had been about one and one-half million dollars, the budget for the fiscal year 1949–50 was over ten million dollars.[20] During this period the parks bureau finally managed to extend its natural parks program into the Maritimes, acquiring Cape Breton Highlands National Park in 1936 and Prince Edward Island National Park in 1937. The budget of the historic sites agency increased proportionally. Whereas in 1935–6 spending on historic sites had amounted to twenty-seven thousand dollars, by the fiscal year 1949–50 Cromarty's operational expenses amounted to almost one million dollars.

The most dramatic turn of events came in June 1938 with the announcement of the federal budget. Although King and his minister of finance, Charles Dunning, were firm believers in a balanced budget and therefore sought to reduce spending, a deepening of the depression in 1937 invited drastic measures. The increasing popularity of the economic theories of John Maynard Keynes which argued for deficit spending as a means of stimulating economic growth coupled with the recent announcement of Roosevelt's programs of massive public spending provided ammunition for the younger members of cabinet who argued for greater government intervention in the economy of the country. Further impetus for a new fiscal policy came from the report of the National Employment Commission tabled in April which argued that federal make work projects were preferable to direct relief and identified the construction and tourist industries as areas in which the government could expand its activity.[21] Gradually the new wave in the cabinet, of which Crerar was a central figure, prevailed, and the 1938 budget was governed by a new set of criteria which placed greater emphasis on spending than restraint in order to encourage economic recovery. Key to this new policy was the national recovery program which provided fifty million dollars for depression relief projects.[22] The national parks bureau received two million dollars of this appropriation of which ninety thousand dollars was designated for historic sites. A further sixty-five thousand dollars was allocated for heritage development by a special vote the following year.

This money far exceeded anything that had been allocated to historic sites in previous years. Plans that had long been in the works to further develop places like Louisbourg, Fort Beauséjour, Fort Chambly, and Fort Wellington were now rushed to completion. At Fort Beauséjour the museum was extended and a caretaker's residence completed at a cost of twenty thousand dollars while at Fort Chambly a long-needed retaining wall to prevent further erosion of the river bank was built at a cost of eleven thousand dollars. Three completely new developments were inaugurated: the Port Royal Habitation near Annapolis Royal, the Sir Wilfrid Laurier House in St Lin, Québec, and Fort Malden in Amherstburg, Ontario. In addition large sums of money were allocated for the restoration of the Halifax Citadel and the old walls of Québec City. This work was to be carried out under the supervision of the Department of National Defence. But it was the three new heritage projects undertaken by the parks service that marked a new departure for the heritage program. For the first time the government seemed willing to spend money to purchase private property for public use and agreed to relatively elaborate development schemes. And, although large sums had also been spent on Louisbourg and Fort Beauséjour, these projects had developed over a number of years while the new schemes were implemented in a matter of months. So while the government's new interest in heritage development was welcome, the haste in which some of the projects were carried out would cause problems for the future.

Once again during this new phase of heritage development the board remained aloof. Partly this reflected its long history of disinterest in projects of this kind. Of the members, only Webster seemed especially devoted to preservation issues and he had long ago come to rely on personal initiative rather than speaking through the board in dealing with these interests. Moreover, like Harkin, by 1938 parks officials had no desire to complicate the allocation of their historic sites appropriation by involving the board. Gibson, the branch director, asked Williamson, the controller of national parks, whether or not the board should be consulted about the spending of the special appropriation. His answer was a rather dispassionate view of the relations between the bureaucracy and the board. He pointed out that his officials had been embarassed by embarking on heritage initiatives and then seeking the approval of the board after the fact. Better, he suggested, not to seek the approval of the board at all and proceed as they had planned.

> I think as far as we could go, in view of the circumstances outlined, is to obtain the co-operation, by way of advice, of individual members of the Board in a case in which he is particularly interested. Such a member would then be acting in his individual capacity and the Department would receive the benefit of his historic knowledge as an individual.
>
> This may be illustrated in the cases of the Halifax Citadel in which Professor Harvey is particularly interested and has the historic knowledge and the local colour of which I think the government might take advantage.
>
> Dr. Webster has been particularly anxious to have Champlain's Habitation restored and I think that we might avail ourselves of his knowledge also in this connection.
>
> I do not recommend that the Historic Sites Board be consulted in any of these works to be undertaken under Supplementary Vote this year beyond this proposed individual consultation.[23]

This statement recognized what was already latent in the attitude of the agency, but the infusion of extraordinary funds had changed the rules of the game, and the board was not even going to be invited to participate.

What did the board members think of this situation? Evidence suggests that they were happy to keep the board apart from what they considered to be political decision making. Howay wrote Cruikshank about the special appropriation in 1938, agreeing that they should have nothing to do with it. "We are going to be loaded down – I mean that Canada is – with little fiddling parochial museums in every place that has a MP, who has the Government's ear or is equipped with great persistency."[24] The other members seem to have complacently accepted the inevitability of government interference and kept their heads down.

Other groups more concerned with preservation issues were caught unaware by the 1938 initiatives. Such was the case with the promoters of the reconstruction of the Port Royal Habitation. The Annapolis Royal Historical Association and the Associates of Port Royal had long been lobbying Ottawa to undertake the development of the habitation but, although they continued their efforts, by 1938 had grown disillusioned about the possibility of a quick decision and were actively pursuing other avenues to achieve their goal. They approached the Nova Scotia government to sponsor the project, stepped up a fund-raising campaign, and in 1937 acquired a small plot of land believed to be the site of Champlain's garden.[25] It was, therefore, something of a surprise when in June 1938 Crerar announced that sixty thousand dollars had been allocated for the development of the Halifax Citadel and the Port Royal Habitation, the amount to be divided equally between the two projects.[26]

However quickly the decision to proceed at Port Royal had been taken, it was inevitable that the local enthusiasts would become involved in the implementation of the project. Although the government appropriation reflected the estimates prepared by the Surveys and Engineering Branch, the location of the original buildings, the plans, and specifications were based on research provided by Harriette Taber Richardson and C.W. Jefferys. The aid of the Annapolis Royal Historical Association was enlisted to acquire the property. Unwilling to wait for the funds to become available, and the necessary order-in-council to be drafted, the department arranged to have the local group purchase the property using funds made available by the provincial government. Then, when the necessary papers were ready, the key lot plus the site of the garden were purchased from the association.

In proceeding with the reconstruction of the habitation there seemed to be general consensus among local enthusiasts, officials, and politicians who advocated for a full-scale replica on the site of the original structure. It would not only be educational, illustrating daily life in an early European settlement in North America, but it would draw tourists as well. Yet there was one small obstacle to this: the Historic Sites and Monuments Board. In 1934 it had been asked to give an opinion on the project to which "it was agreed that no action by the Board should be taken with regard to the proposal ... to erect a replica of the French stronghold built by Champlain in 1605."[27] Despite the enthusiasm of Webster, and later, Surveyer, the others were resolutely opposed to the idea. Cruikshank seemed to summarize the position of the majority when he wrote: "in my opinion these attempts to reconstruct buildings which have entirely disappeared and are only known from vague descriptions or plans of doubtful authenticity with modern materials and workmen of the present time are absurd and a mere waste of money."[28]

The board's reservations about reconstruction being ahistorical did not merely reflect an aversion to heritage development, it represented a particular perspective toward the conservation of historic property. Heritage development involves contradictory aims: it must preserve the historical integrity of the site while providing means for the public to enjoy and understand it. The aim of preservation is to keep the site in its original state, unspoiled by modern encroachments and protected from further natural deterioration. Yet in order to aid the public's appreciation of historic remains, some intervention is usually necessary. Roads and fences need to be built and museums or interpretive centres established to introduce the development. Often the structures themselves are altered to enhance their historical interest. Buildings that have fallen down or have undergone alteration in subsequent historic periods invite treatment to restore their appearance to that of a bygone age. What treatment is selected depends on a number of factors. There are fashions in restoration philosophy just as there are in other sciences. In the late nineteenth century Ruskin's dictum that any restoration was destructive had considerable influence. Later American trends toward historical reconstruction, culminating in the replication of colonial Williamsburg in the 1930s, represented another pole of restoration philosophy. Today restoration is considered to be acceptable so long as replicated elements are distinguished from the original structure. Total reconstruction is not considered to be preservation.[29]

Despite its limited means, the parks service had acquired some considerable experience with preservation techniques. Although influenced by the views of Adams, the town planner, that modern restoration was to be avoided, it had countenanced some minimal intervention, as at Louisbourg and Fort Beauséjour, to rebuild fallen walls and excavate filled-in trenches. This process was taken a step further at Prince of Wales's Fort where a missing element, the gateway, was reconstructed with modern materials to provide a sense of structural unity to the fortification. Even the board approved of these preservation efforts and, although modern restoration architects might not give their imprimatur, they were not destructive in any real sense.

Total reconstruction was just carrying this process a stage further, yet at this point ideals of preservation had been left far behind in an effort to interpret an historic site. While the building of a modern structure may enhance the interpretation of the site, it contravened a basic aim of preservation. Any original remains would be lost forever. Although not widely held at the time, the board strongly believed that reconstruction was contrary to preservation. Unfortunately the board was alienated from the preservation part of the heritage program, and so an important point of view was ignored.

This is not to say that the local enthusiasts and federal officials involved with Port Royal were not concerned with authenticity. They were. Mrs Richardson, C.W. Jefferys, and others had done painstaking research to ensure that their facts were right.[30] Col E.K. Eaton, who had succeeded L.M. Fortier as superintendent of Fort Anne, asked Dr C.T. Currelly, an archaeologist with the Royal Ontario Museum, to act as consultant to the project. Although Currelly agreed, delays in beginning the excavation prevented him from becoming involved so one of the Associates of Port Royal, a Professor Pond of Harvard University, proposed C. Coatsworth Pinkney to undertake the archaeology of the site.[31] Pinkney was a landscape architect who had restoration experience from Mount Vernon and Williamsburg and some knowledge of modern archaeological methods. As the American associates offered to pay his salary, Ottawa readily agreed to his involvement. Pinkney started work in October 1938, sifting the soil on the site to a depth of twenty-one inches, uncovering old foundations, and gathering artifacts. His findings helped the architect determine the exact dimensions and location of some of the buildings.

The architect in charge was K.D. Harris, from the surveys and mapping branch, who had previously worked on the construction of Fort Anne. In 1935 the branch had taken the unusual step of replacing the Fort Anne Officers' Quarters with a concrete replica. Harris was assisted by C.W. Jefferys who had prepared plans and elevations based on drawings in Champlain's narrative. Further details were obtained from a French architect, an expert on old buildings of Normandy and Picardy, and Ramsay Traquair who provided information on building techniques in New France. Work carried out by local craftsmen during 1939 attempted to simulate an antique appearance: timbers were finished by hand and stone dressed to look suitably weathered. During the 1940s about forty pieces of furniture were made at the Acadia Forest Experiment Station to designs prepared by Jefferys and placed in various rooms of the habitation.[32] Apparently Marius Barbeau of the national museum was also commissioned to furnish pieces for the interior and to draw plans for the interior design of the chapel.[33] Cromarty tried to ensure that all pieces were of the proper period.

Despite this commitment to authenticity, problems emerged in the attempt to create an exact copy. Champlain's illustration did not depict all of the buildings, and the appearance of some were based on conjecture. The planners had only vague notions about the original construction methods. They assumed, for example, that the walls of the buildings had been *pièce sur pièce*, that is, with vertical timbers placed on a horizontal sill. Recently, however, it has been suggested that a more likely method was *colombage*, where the vertical timbers were spaced and the gaps filled with wattle or stone.[34] The planners also overlooked details in replicating building materials. A plaque erected over the entrance, for example, not only depicted the wrong coat of arms for the de Monts family, but was fashioned from plywood.[35]

Problems also arose in the interior reconstruction. Although the participation of Jefferys and Barbeau ensured a high level of authenticity, the banquet room, the famous setting for the Order of Good Cheer, became the exclusive preserve of the Annapolis Royal Historical Society whose enthusiasm sometimes glossed over obvious anachronisms. A musuem planned by the local group tended to be eclectic rather than focused on the history of the site. These transgressions attracted the ire of Harvey who got more involved as Webster became more frail. Writing to Webster in 1947 be complained "[a]s you know, I agree absolutely that something must be done to guide, if not control, the egotistical efforts of Eaton and Merkel, neither of whom has as much concern for historical accuracy as for personal publicity."[36] These concerns led to Harvey's participation on a management committee, along with Harris and Cromarty, to maintain proper standards. And, despite these occasional lapses, there was general consensus that the reconstruction provided a meaningful educational experience about early Canadian life and architecture.

More serious problems arose later, however, when subsequent investigations failed to confirm that the habitation was in fact on the original site. Archaeological investigations conducted in the mid 1960s raised some doubts about the authenticity of the site.

Moreover, it was realized that the original excavations had possibly destroyed important clues. In the words of the staff archaeologist: "As far as I have been able to determine, there is no full report on the excavations. We have the artifacts, some photographs and some generally uninformative progress reports. In short we have nothing which could be construed as an archaeological report in terms of present day standards ... My tentative conclusion is that Pinckney [*sic*] did not excavate the Habitation. Rather I believe that he assumed that he was digging on the correct site and simply interpreted everything he found as belonging on the Habitation."[37] The habitation was one of the showpieces of the historic park system and helped pave the way for more elaborate developments in the future. But preservation it was not.

Principles of heritage conservation were likewise compromised in the development of the birthplace of Sir Wilfrid Laurier in St Lin, Québec. A monument had been erected in St Lin by the parks branch in 1927 following the initiatives of the National Committee for the Celebration of the Diamond Jubilee of Confederation. The plaque was sponsored not by the board but by the national committee, which seems to have had quasi-official status and which also recommended commemorating the site of Sir John A. Macdonald's grave in Kingston. Nonetheless, both sites came to be regarded as *bona fide* national historic sites. Although another plaque had been erected across the street from Laurier's birthplace in 1925, no effort was made to identify any particular house associated with Laurier's birth, and no further action seemed necessary.

Then in 1934 a Montréaler, Arthur Christin, purchased a house in St Lin which he claimed to be the birthplace of the former prime minister. Following the election of the Liberal government the next year, high-level recommendations were made to acquire the house for development as a national shrine, and in 1937 Crerar directed the parks branch to report on the building's potential. The chief inspector of parks, James Smart, visited the house in September and noted that "[t]he building appeared in good repair, but is not a very imposing structure."[38] At the same time the department moved to acquire the property. The board was not consulted about the site until November when Williamson telephoned the members to ask if they would object to the property being developed as a national historic site. Although they unanimously agreed, no one took any particular notice of the project.[39]

The government purchased the property in December, and the board gave its formal approval the following May. At this point the parks bureau was deeply committed to preserving an historic site about which it knew very little and had little idea of how to develop. It had only hearsay evidence that the house was indeed the former Laurier residence and it had no plans and no budget with which to proceed. But many things happened in 1938.

In January the Montréal landscape architect, Frederick G. Todd, offered his services in devising a development scheme. Although parks officials had no indication that money for the development was forthcoming, they were told by the minister to accept Todd's free advice. Todd's plans, which were received in May and August, reflected a particular approach to heritage development. He had been the landscape architect in charge of the development of the Plains of Abraham. There he had been principally concerned with landscaping the grounds to make an impressive visual statement rather than interpreting historical themes. This concern also guided his approach to the development of the St. Lin property. He proposed buying up the surrounding lots so that unsightly buildings could be removed, and relocating the Laurier house to the centre of a park-like setting to make a "proper memorial."[40]

There were some reservations about Todd's proposal. G.W. Bryan wrote Williamson in September saying, "Mr. Todd's letter would indicate that he has in mind the establishing of a park on the Monahan [i.e. the neighbouring lot] and Departmental property. Such a plan would not be in keeping with the appearance of the area when Sir Wilfrid lived on the residence now owned by the Department and would be contrary to the general conception that improvements to historical sites should not be modernized."[41] But Bryan was overruled, and the department began negotiating to acquire the neighbouring property to carry out Todd's proposal.

Meanwhile, the department commissioned Marius Barbeau to report on the site. Barbeau visited St Lin late in 1938 and submitted his report in December. He confirmed that the house was the birthplace of Laurier but noted that it had subsequently been moved forward on the lot. Since its relocation had already affected its historic integrity, he felt there was no reason why it should not be moved again, and, in light of difficulties in obtaining the neighbouring lot, he proposed moving the house across the street beside the old public school once attended by Laurier. Todd concurred with this suggestion: "[i]f Dr. Barbeau feels that the historical side of the question will allow of the removal of the cottage to the school property you indeed have an ideal solution of the problem and I congratulate you and Dr Barbeau. Certainly I could not think of any means of making a suitable memorial to Sir Wilfrid and retain the house in its present location."[42] But the department continued to negotiate for the purchase of the Monahan property which it finally acquired in February 1939.

By this time the department had a special allocation of four thousand dollars with which to develop the site, and work was carried on in the ensuing months.[43] The house was moved to the centre of the two lots, giving it a more dignified setting, and allowing for the construction of a new concrete foundation. The walls of the house were renovated and the grounds landscaped. The caretaker was authorized to purchase furnishings to restore the interior to its supposed appearance at the time of Laurier's birth. In July 1940 it was reported that "[t]he furnishing of the building is complete and exceptionally good taste has been exercised in securing furnishing ... which are of the French habitant style of about the time of Sir Wilfrid's birth."[44]

The Laurier house does not seem to have been taken seriously as a preservation work at the time, being mainly regarded as a shrine to a great Canadian. It is taken even less seriously today as a work of preservation as it contravenes a number of basic principles. A thorough historical investigation was not conducted until the 1970s and this tended to confirm rather than deny a persistent fear that the house had never been inhabited by Laurier.[45] Subsequently the site has been renamed Parc national Sir Wilfrid Laurier, evading the question of whether the house was actually Laurier's or not.

But what does it matter if the house is the real thing or not? Like the Port Royal Habitation, it served the objectives of the parks bureau to represent a way of life in a particular time and place. Despite anachronisms, both manage to provide information through buildings and artifacts about segments of the Canadian past. In this context they can be considered to be more like museums than exercises in historic conservation. This attitude toward heritage development was reflected in the tendency of the parks bureau in this period to refer to its heritage projects as historic museums. And it governed a third major initiative in this period, the acquisition and development of Fort Malden in Amherstburg, Ontario.

Local enthusiasts for the preservation of Fort Malden had formed one of the lobbies before the First World War that had led the government to establish a heritage agency in 1919. But although the board recognized the fort as a national historic site in 1921, Cruikshank had argued against further involvement. He wrote Harkin on the subject in 1935: "[a]s you are aware, my opinion is that the creation of small local museums at more or less out-of-the way points, which have certain historic interest, is a mistake, and to a certain extent a waste of public money."[46] The major drawback to its preservation was that there was little to preserve. The fort had been largely destroyed, and the land appropriated for a housing development. Still, the local enthusiasts persisted, and when the real estate scheme fell through, it renewed its pressure on Ottawa to preserve the site. Local interests coalesced in the 1930s through the organization of the Amherstburg Historical Sites and Museum Association led by Major A.W. McNally.[47] In the early 1930s the property was acquired by the municipality in lieu of unpaid taxes, and through the efforts of the association in 1937 it was transferred to the federal government to be administered by the parks bureau as an historic site.

Although the board repeated its recommendation against further development, in 1937 the local group, now organized as the Fort Malden Management Committee, extracted a more tangible commitment from Ottawa. In this endeavour they were fortunate to coincide with the announcement of the bounties of the national recovery program. The parks bureau agreed to build a museum, and Cromarty designed one along the lines of his French Château buildings at Louisbourg and Fort Beauséjour.[48] Its operation, however, was controlled by the Fort Malden Management Committee. It appointed a full-time curator in 1941 and supervised the collection of artifacts and curiosities which were displayed in military, pioneer, and Indian rooms. In 1946 the government acquired an adjacent piece of real estate containing some desperately needed heritage architecture, and the site finally boasted a brick barracks, bakery, and laundry from the old fort besides the modern museum building.[49]

In 1939 parks officials began exploring means to rationalize these developments. They had never been very happy with the inclusion of heritage areas in the national park system, and the 1930 legislation had provided for a distinct category of national historic park. But it was not until 1939 that Williamson and Cromarty decided to implement this part of the National Parks Act by classifying a number of historic sites and national parks as national historic parks. This raised the problem of definition, and the parks staff pondered various sorts of criteria before deciding that all sites which were extensive enough in size would be included in the new category.[50] Consequently, in 1940 an order-in-council was passed bestowing national historic park status on the Fortress of Louisbourg, Port Royal Habitation, Fort Anne, Fort Beauséjour,

Fort Lennox, Fort Chambly, Fort Wellington, Fort Malden, and Prince of Wales's Fort. No one seems to have minded the unbalanced nature of the category. Only Port Royal Habitation was a nonmilitary structure and the only archaeological site controlled by the department, Southwold Earthworks, languished in the cold. Nonetheless, the implementation of the historic park category gave the department better control over its expensive heritage operations. As national historic parks these properties would receive a regular annual appropriation and be maintained by a salaried superintendent directly responsible to the parks bureau.

CHAPTER SIX

The Board in Familiar Waters, 1934–54

During the 1930s the board steadily lost ground in its ability to influence government policy. Although it expressed definite views about what not to undertake in the way of preservation, it had few positive suggestions for heritage development. In an era when the department wanted to spend money on heritage projects this meant that its usefulness as an advisory body was seriously curtailed. But it was not just over the issue of preservation that the board was alienated from government policy making. With the growing complexity of government, there were other experts within the bureaucracy advising on heritage development. The lines of communication were much more diffuse in the late 1930s than they had been in the early 1920s.

The board had lost an important conduit to parks policy making with the retirement of Harkin. Although Williamson replaced him on the board, the controller did not participate in its activities to the extent that his predecessor had. Following Williamson's death in 1940, the new controller, James Smart, did not join the board, and left all liaison duties to Cromarty who continued on as secretary. Although relations between Cromarty and the board seemed amicable, this situation served to increase the bureau's hold on preservation work and widen the gap between the two halves of the heritage program.

An indication of the board's waning influence on departmental decision making was the way in which two new members were appointed to the board in 1938. Father Antoine D'Eschambault was named to represent Manitoba, and J.A. Gregory, Saskatchewan. Although Howay was friendly with D'Eschambault and approved of his appointment, he had not been consulted about either nomination, nor, it would seem, had any of the other members. Howay did not approve of the appointment of Gregory and complained to Webster about him. "For my part I am rather disgusted (whisper it not in Gath) at the inclusion of Mr. Gregory MP He is not on the map in matters historical, and the appointment smacks overmuch of politics."[1] Cruikshank believed that the new members would complicate matters as they would bring new interests to a program whose objectives he felt had largely been fulfilled.[2] Webster considered the appointments to be an insult to Howay and spoke darkly of "Politics! Politics!"[3] Harvey, too, thought that politics were behind the appointments. "I think perhaps the fact that the Hon. T.A. Crerar comes from the West accounts for it, but I don't like to see the balancing of a Frenchman and an Englishman in the West as in the East. I shouldn't be surprised if a Frenchman were added for the Maritime Provinces."[4]

The new appointment made the board feel uneasy about its ability to control its destiny. Although differences existed among the older members, there was enough similarity of background to give a clublike atmosphere to the board's proceedings. Now that outsiders were being allowed in, who could say what else the government might do? Harvey was particularly paranoid about political interference and his suspicions increased following the death of General Cruikshank in 1939. Ottawa decided not to give Ontario another representative and instead appointed Dominion Archivist Gustave Lanctôt to the board as an *ex officio* member, leaving Fred Landon to represent all of Ontario. Lanctôt gave the French even greater representation on the board. Following the death of Howay in 1943, who had succeeded Cruikshank as chairman, Harvey, determined to block a francophone appointment to the chair, argued that Webster should be chairman on the grounds of seniority. "If not," he warned Webster, "I imagine that the politically-minded Lanctot will be getting in his word on behalf of a French-Canadian chairman, and as Surveyer would be impossible he himself would aspire to the position."[5]

While the appointment of the new members represented the political weakness of the board, the new members did not immediately exert a great influence on its proceedings. D'Eschambault (1896–1960) was an assistant to the archbishop in St Boniface and active in the historical society there. Interested in the early history of the west, he later became a recognized historian, being elected to the Royal Society in 1954. In 1960, the year of his death, he was elected chairman of the Historic Sites and Monuments Board. His early years on the board, however, were relatively quiet and he had little influence on new initiatives until after the war. J.A.

Gregory (1878–1950) exerted even less influence. A provincial MLA, he was also president of the Prince Albert Historical Association. His participation on the board was largely confined to passing along the suggestions of his historical society. Even this function largely ended following his election to the Canadian Parliament in 1940 when he took up residence in Ottawa. Then a prolonged illness leading up to his death in 1950 precluded all participation on the board.

Two stronger appointments were made in 1944 with the nomination of Walter N. Sage from Vancouver to replace the deceased Howay, and M.H. Long from Edmonton to represent Alberta. Both were senior professors of history at their respective provincial universities and both were active in local historical societies. Moreover, both members became extremely diligent in representing their provincial constituencies, traveling extensively in the summer months to investigate possible historic sites and co-ordinating heritage efforts among the local, provincial, and national levels. Sage in particular, although personally disliked by Howay, seemed to have inherited much of the judge's influence in provincial heritage circles.

Unfortunately, Sage and Long tended to confirm rather than alter the board's earlier direction. Both were more interested in commemoration rather than preservation, perhaps reflecting an academic bias against material history. Even in the context of academic history their views were distinctly old fashioned, placing them closer to the ideas of Cruikshank, Howay, and Webster than the new generation of professional historians. Lewis H. Thomas noted in his obituary of Long, for example, that "rooted in his disposition and university training were Professor Long's abiding admiration for British ideals of civilization and his concern for preserving a meaningful Canadian Association with Britain and the Commonwealth."[6] These views, shared by Sage, Harvey, Landon, and Webster, gave further impetus to the nationalizing mission of the board's commemorative work.

Cut off from the preservation side of the heritage program and reinforced by new members more concerned with commemoration, the board concentrated on its task of erecting plaques to explain national history to the public. During this period it strove to widen the range of its commemorations, adding new topics and raising the number of designations in under-represented regions.

The board took a stride in this direction in 1937 when it finally agreed on a means of recognizing historic figures through secondary plaques. This allowed the board to expand into what it had previously felt to be problem areas of national significance such as arts and letters. Anyone of sufficient fame would be considered to be worthy of a secondary tablet including provincial premiers, painters, poets, and popular novelists. This initiative provided greater opportunity for local concerns and special interests to gain representation in the national pantheon. Thus historical events outside of the mainstream tradition of French- and English-Canadian national development gained a modicum of recognition. In 1939, for example, the board approved the commemoration by a secondary marker of Louise Crummy McKenney, the first woman to be elected to any parliament in the British Empire.[7]

Because secondary tablets were excluded from the normal quota of commemorations allocated annually to each region, they offered an opportunity for other regions to catch up to the larger number already marked in Ontario. Harvey saw secondary tablets as a way of redressing the imbalance in the distribution of monuments. Writing to Webster in 1939, he said: "I myself have learned that the Maritime Provinces as a whole have fewer tablets than Ontario alone. I think that in the next two years you and I should try to catch up in this respect. The only way I see to do it is through the secondary tablets to the Fathers of Confederation and other distinguished men born in the Maritime Provinces"[8] This campaign paid dividends for the Maritime members as the initial recommendations for secondary markers were heavily weighted in favour of the eastern provinces. Prince Edward Island, especially, made signficant gains in this way. Subsequently, the distinction between primary and secondary sites was lost, and the secondary monuments came to represent fully-fledged national historic sites.

Secondary tablets were successful in broadening the scope of the board because they avoided the question of national significance. But this problem still complicated the designation of regular historic sites where different regional and ideological perspectives on national identity sometimes led to conflict. As might be expected, the Québec point of view was frequently at odds with those of other regions, and even the diffidence of Surveyer was sometimes provoked to outrage when the rest of the board refused to recognize Québec historical concerns. At the 1939 meeting of the board, for example, Surveyer, acting on the recommendation of a heritage group in Rimouski proposed commemorating the Acadians at Bonaventure. When the rest of the board turned him down, Surveyer was furious at least partly, one supposes, because of his loss of face in the Québec historical community. He thought it particularly unfair that an Acadian site had been rejected when loyalist sites had been commemorated in the Maritimes. He then railed against the Ontario bias of the designations. "If we consider that Quebec has had so far 46 monuments or tablets as against Ontario's 71 with less history behind it, I cannot help feeling that the standard of national importance is not the same for Quebec as for other parts of the country, and I do not feel inclined to submit propositions for the sake of having them turned down."[9] Yet while Judge Surveyer complained about the narrowness of his colleagues, he was himself guilty of a parochial outlook. He regularly turned down requests to recognize Jewish historical sites in the province, and when Judge Howay proposed a monument in British Columbia to commemorate the 400 black Americans who came to Vancouver Island before 1858, he wrote, "I do not think the immigration of negroes is a fact to rejoice upon."[10]

Similar narrow concerns deterred the commemoration of broader historical themes in the west, despite the addition of new members to represent the western provinces. Although Alberta historical societies pressed for the preservation of a buffalo pound, Howay did not wish to acquire a site that he considered to represent the extinction of the buffalo.[11] Father D'Eschambault did not regard Mennonite immigration as a suitable theme for commemoration, and a proposal to designate the first Ukrainian settlement in Canada was likewise turned down by the board.[12]

Competing regional perspectives sometimes made it difficult to choose a single site to commemorate themes that transcended regional boundaries. One such case, suggested by a Québec resident in 1938, was a proposed monument to the lumberman. While the board was favourable to the idea, it discovered that a number of regions had legitimate claims to the monument, principally New Brunswick, the Ottawa Valley, and British Columbia. Moreover, these regions had distinct lumbering histories embodying different traditions. How could a single plaque recognize these regional particularities in a national context? A committee struck to investigate the problem recommended that three monuments be erected, one in Atlantic Canada, one in central Canada, and one in British Columbia. The board agreed to this proposal and specified that the inscriptions "should emphasize the importance of the lumber industry but each should note special characteristics of the respective region."[13] This was a rare example of the multiple designation of a single theme. A subsequent suggestion by Professor Sage to have multiple commemorations of the trans-Canada canoe route used by the fur trade was deferred by the board.[14]

Ideological differences also hindered action on important issues. An example was the debate which took place in 1940 over whether the government should finance the repair of the Patriote's monument in Montréal. The original monument had been erected in 1852 by a "rouge" organization, the Institut Canadien, to commemorate the Patriote's who were killed in the rebellions of 1837-8. Now it was in considerable need of repair, and the federal government was lobbied to provide the necessary funds.[15] The question was referred to the board for its opinion. If Cruikshank had still been chairman, the proposal might have been rejected outright, but the ideological balance had shifted, if only slightly, in recent years so that opposing views had greater weight.

One side saw the monument as an important symbol of the development of Canadian nationhood. This was not just a French-Canadian notion but one shared by the liberal members of the board. Both Surveyer, Landon, and D'Eschambault believed that the government should assume responsibility for the monument. They were joined by Howay, who wrote that "whether the actions of Mackenzie and Papineau were well or ill advised is beyond the question, at any rate they thereby, even if guilty of treason, brought to the attention of Home authorities the conditions in Canada with the result that we have the Durham report."[16] But the Maritime members were resolutely opposed to Ottawa's recognition of the monument. Webster argued that to give funds for its repair would be an endorsement by the government of armed rebellion which is "not in keeping with the democratic way of life."[17] Harvey, one of the last of the imperialists, wrote: "a glance at the inscriptions will show that for the Dominion government to take over and repair this memorial would be tantamount to vindicating the action of the rebels and criticizing violently the action of the Imperial government at that time."[18] Fortunately, the cessation of the board's activities because of the war relieved it from having to render a split decision.

Despite limitations imposed by region and ideology, the board in this latter period of its work managed to designate a number of sites that represented other historical themes besides battles, Loyalists, and the fur trade. This was especially true in the regions west of Québec. Ontario acquired a number of nonmilitary sites, commemorating such topics as lighthouses, cheese factories, oil wells, and Indian treaties. The prairie provinces increased their number of sites dramatically and introduced new themes such as the founding of the province, Indian leaders, and the saving of the buffalo. Still, well-marked themes continued to receive attention during this period. Traditional topics were memorialized, and in 1938, for example, the Battle of Lundy's Lane received three tablets.

In this period, the board did not widen its scope to recognize a national architectural heritage. Although the board had regularly turned down requests to recognize historic houses in earlier periods, it came under increasing pressure, some of which was political, to recognize period architecture. The board was lobbied to acquire the Perry Borden House in Grand Pré, Nova Scotia "and to fit up within it a museum containing furniture, records, etc. of the New England Planter."[19] Similarly, it was asked to undertake the preservation of an eighteenth-century merchant's house in Québec City.[20] In Ontario, increased awareness of the plight of loyalist houses inspired an all out campaign to urge Ottawa to preserve some of these buildings. Particular attention was given the former home of Sir Francis Baby near Windsor. A committee was formed to organize its preservation, and plans were drawn up for its construction. But when the local MP asked Williamson why the parks bureau could not sponsor the preservation, he hid behind the stated policy of the board in excusing his inability to act: "all activities along this line are restricted to those sites which are considered by the Historic Sites and Monuments Board of Canada as being of national importance from an historical standpoint."[21] The board did not seem interested in old buildings. Its members found themselves being dragged off to look at many old building in this period, however, and even Cruikshank consented to visit an old log house on the Rideau Canal in 1938. But Cruikshank only had eyes for forts, and while he dismissed the importance of the house, he discovered a "genuine stone blockhouse, in good condition, built before 1834."[22] But there were already many military buildings being cared for as national historic sites, and too few examples of domestic or commercial architecture.

The virtual suspension of the board's activities during the war provided an opportunity for taking stock. Following Cruikshank's death, Howay, as interim chairman, consulted with Webster about the past and future direction of the board. Then, when the board reconvened in 1943, Howay presented a lengthy report on past commemorations and proposed greater expansion of its work. Of 285 national historic sites, he found that 105 commemorated battles and war, 52 exploration, discovery, and the fur trade, 43 illustrious men, and 36 commerce and general development. Only 4 marked outstanding political events, 7 social services and 3 Indians. Designations were further characterized by a regional imbalance. Of

305 sites designated by the board, Prince Edward Island had 13, Nova Scotia, 38, New Brunswick, 31, Québec, 63, Ontario, 97, Manitoba, 15, Saskatchewan, 8, Alberta, 17, British Columbia, 22 and the Yukon, 1.[23]

Howay proposed that the board adopt broader criteria to further enlarge the scope of its designations. "Outstanding events connected with Canada's economic, social, and cultural growth, or with her basic industries: mines, fishing, lumbering, agriculture, horticulture, cattle raising, wildlife deserving commemoration may have possibly been overlooked."[24] And in a surprise move Howay also suggested that distinctive examples of Canadian architecture be commemorated and urged the preservation of typical examples of manors, farm houses, fur trade posts, and mills.[25] This was a new stance for the board to take. It had done more than enlarge its horizons – it had shaken off a numbing timidity that had pervaded the earlier years of its existence.

There was one other episode at this time which, while heralding a new order of government activity in cultural affairs, also looked back to previous development. Coming at the time of a profound reorientation of government policy, the Massey Commission has been credited with ushering in this new era. But as an examination of previous cultural policy, its report is an important appraisal of the history of government cultural agencies until 1950. It relates to the Historic Sites and Monuments Board because the Massey Commission mistakenly assumed that it was central to the operation of the federal heritage program. Like the board, its outlook is also characterized by a curious mixture of old- and new-fashioned ideas about the role of the state in promoting cultural identity. Thus, the report of the Massey Commission sits, Janus-like, at the juncture of two eras. While the report influenced the subsequent administration of the national historic sites program, it was the last attempt to understand the problems of the old order.

The Royal Commission on National Development in the Arts, Letters and Sciences was formed in April 1949 with the following terms of reference: "That it is desirable that the Canadian people should know as much as possible about their country, its history and traditions, and about their national life and common achievements; [and] [t]hat it is in the national interest to give encouragement to institutions which express national feeling, promote common understanding and add to the variety and richness of Canadian life, rural as well as urban."[26] The commission was directed to investigate the role of federal agencies in achieving these goals and recommend ways for improvement. The institutions singled out for attention were the Canadian Broadcasting Corporation, the National Film Board, the National Gallery, the National Museum, the National War Museum, the Public Archives, the National Library, and the Library of Parliament. The commission was also directed to report on Canadian involvement in UNESCO and "relations of the government of Canada and any of its agencies with various national voluntary bodies operating in the field with which this enquiring will be concerned."[27] The heritage program of the national parks service was not initially identified for investigation and it only belatedly came under the purview of the commission.

The leading member of the commission, and its chairman, Vincent Massey, had been instrumental in bringing about its formation. Involved in the larger Canadian cultural scene through his work with the Massey Foundation and at Hart House at the University of Toronto, he was a leading patron of the arts. He had long been active in the Liberal party and during the 1930s had led a movement within the party for a more interventionist public policy.[28] He had inspired a younger generation of Liberal thinkers which, combined with issues such as the question of a national broadcasting policy, helped convince Prime Minister St Laurent of the need for a royal commission to investigate all aspects of cultural policy.[29] Massey helped choose the members of his commission. These were: Arthur Surveyer, a Montréal engineer and brother of board member, Edouard-Fabre, Rev. Georges-Henri Lévesque, dean of the faculty of social sciences at Laval, Hilda Neatby, professor of history at the University of Saskatchewan, and Norman Mackenzie, president of the University of British Columbia. Massey seemed to work most closely with Lévesque and Neatby, and the trio was responsible for most of the commission's recommendations.[30]

The commission's findings were incorporated in the published report of the Massey Commission, presented in June 1951. The report consists of two parts: the first summarizes its findings in the various fields of inquiry, and the second presents detailed recommendations on future cultural policy. There is a remarkable amount of optimism reflected in the report for, although its findings painted a bleak picture of the precarious state of Canadian culture, implicit in many of its comments was the belief that once the government took charge and provided adequate funding and a comprehensive policy, then the situation would change dramatically for the better. As a detailed examination of the state of Canadian arts and letters and as the blueprint for greater government intervention in cultural development, the Massey Commission report is one of the most significant government documents of the postwar era. It was instrumental in shaping future communications policy and, after a prolonged wait, the Canada Council. To a lesser but still significant extent, the report also influenced the future direction of the historic sites program.

In analyzing historic sites, however, the Massey Commission incorporated the same ambivalence about national significance that had characterized the earlier work of the heritage program. On the one hand, Massey and his colleagues were custodians of culture, convinced of the need to identify a unifying national heritage. In this way they identified with the commemorative work of the Historic Sites and Monuments Board. But the Massey Commission was also committed to presenting the views of special interest groups, and these brought a different perspective to the work of the historic sites program. Presentations heard at regional hearings complained about the uneven distribution of sites across the country and emphasized the regional nature of national history. They also focused on the need to make more effort in the area of preservation, especially with regard to domestic architecture, an aspect that had little interest to the custodians of culture on the Massey Commission.

These views fuelled a contradiction in the report of the Massey Commission for they led to two sets of recommendations which were not altogether in harmony. On the one hand, there was the perspective of the commission's members such as Massey and Hilda Neatby, who favoured a strong central agency imposing a common cultural tradition on a disparate society. As custodians of culture, they preferred commemoration and, like Cruikshank in the past, regarded favourably the preservation of military sites. On the other hand, the Massey Commission actively sought the views of local groups concerned with heritage issues. These groups favoured greater regional expression within the national pantheon of historic sites and placed greater emphasis on architectural preservation. These two perspectives determined the nature of parts of the final report. The first part of the report on historic sites was largely concerned with passing along complaints received at the regional hearings. Not surprisingly, then, the criticisms centered on the unevenness of the program and its central Canadian bias. It noted that while Ontario had 119 out of 388 commemorated national historic sites, Saskatchewan had only 8.[31] It repeated the criticism voiced by Saskatchewan historical societies that not enough attention was paid to the prairie provinces, and that greater emphasis needed to be placed on prehistoric sites. It also voiced the complaint of a number of historical societies that the board was out of touch. "Another cause of serious complaint, from Saskatchewan particularly, but also from Quebec, is the failure of the Board to get in touch with other interest groups, to explain its policy, to keep them informed of its activities, and generally to agree on a proper division of interest." The commission also repeated a recommendation of the Royal Architectural Institute of Canada which urged "the preservation of old houses of architectural merit while ... [noting] that there was a tendency to favour military sites and that most of the restoration projects involved forts."[32]

At the crux of the commission's recommendations for change was the call for a clearly stated policy on the marking and preservation of historic sites. One of its leading recommendations, then, was that the board "undertake a much more comprehensive program in the future and that it be provided with funds adequate for its important reponsibilities."[33] The board was also charged with developing a national policy on preservation and placing greater emphasis "on the restoration and preservation of historic sites and buildings including those buildings of purely architectural significance." The board was also directed to improve its commemorative program, especially in regard to the design of the plaques and commemorative cairns. The board was further directed to co-operate with provincial heritage programs and "act as a clearing house for information."[34]

While the commission viewed the Historic Sites and Monuments Board as being largely responsible for the federal heritage policy, it conceded that structural changes would be needed if it was to carry out its recommendations. It therefore proposed its reorganization, providing for greater representation from central Canada. Ontario and Québec would have two representatives each, while the other provinces would continue to have only one. Moreover, it recommended that two additional members be appointed on the recommendation of the Canadian Historical Association. The dominion archivist would continue to serve *ex officio*. The commission proposed that the secretary to the board should be a civil servant who was also "a professional historian of established reputation,"[35] thereby ensuring competent direction in the parks branch.

The report did not completely overlook the role of the parks branch in implementing historic sites policy and made several recommendations pertaining to its organization. Adequate funds were to be made available for the preservation agency, and in particular, funds were to be made available for the preservation of the Halifax Citadel as a heritage property. Ways and means were to be devised so that other heritage properties currently controlled by the Department of National Defence could be transferred to the care of the national parks service. As well, although it was realized that the federal government had no constitutional right to legislate to preserve privately owned property, the report urged that the government "suggest to the Provincial Governments that they take suitable legislative action to protect historic sites and buildings by scheduling them in the national interest as is done in Great Britain and in France."[36]

The report inspired a long overdue re-evaluation of the direction of the historic sites program by the government. Following the publication of the Royal Commission's report, the minister of Mines and Technical Resources sought the reaction of the Historic Sites and Monuments Board. Members were invited to submit a written memorandum which was to be forwarded to the board's chairman, Fred Landon. Landon incorporated the various comments into a single discussion paper which was tabled at the May 1952 meeting of the board. The meeting sought a consensus from the diverse views originally expressed by the members and produced a set of recommendations to counter those of the Massey Commission.

Predictably, the initial reaction of individual members to the report was defensive, and there was some bitterness that their work had not resulted in a more favourable report card. At this juncture there were a few members who had given years of service, as well as newer arrivals who had been extremely conscientious in their duties as provincial representatives. While this experience justified a certain amount of collective self-pity, it also allowed some members to make cogent rebuttals to the commission's analysis as well as provide informed insights of their own into the program's shortcomings.

Harvey, for instance, noted the inconsistency between the report's criticism of a central Canadian bias and its recommendation that Québec and Ontario be given additional representation.[37] It was also obvious to the board that the criticism that there was too much emphasis on military sites was not consistent with the recommendation that the parks service preserve more forts. Most of the members agreed that, in the recent past at least, allocation of sites had been made without overt regional bias, and that the limitations of the present program were largely due to inadequate funding. While the board recognized that the historical themes were unevenly represented, Landon argued that this was due to the evolving nature of the program.

It was natural, I think, that much attention would be given in the earlier period to the older provinces and to the military features, particularly the War of 1812, since the first chairman was this country's greatest authority on the struggle with the United States. But this Board has since gone beyond battlefields and forts and in recent years the influence of social history upon its work has been marked. We have long been commemorating peacetime achievements and the cultural side of Canadian life, as well as honouring the achievements and records of individual Canadians.[38]

This was a telling point. How could the present organization be responsible for a body of commemorations that had evolved over thirty years? The board's reaction to the charge that it had failed to reflect the concerns of local heritage groups was less sure. The members had worked long and hard to bring the views of the Maritimes, Ontario, and British Columbia into the national forum, but it was vulnerable to criticism that it had ignored Québec and Saskatchewan. The member representing Saskatchewan, from where the bulk of public criticism had stemmed, had died in 1950. The new member, Campbell Innis, was not in a position to defend a record for which he had not been responsible. Judge Surveyer refused to answer the charges initiated at the Québec hearings on the grounds that the commission had not undertaken a serious inquiry into the board's work. Walter Sage, however, in a confidential note to the department, suggested that it was the past inadequacies of the Québec and Saskatchewan members, not structural faults of the board, that had provoked the bulk of public criticism.[39]

The board's recommendations for reform were consistent with its rebuttal of the commission's critique. It agreed that it should broaden the scope of its activities and do more in the way of preservation, provided more adequate funding was forthcoming. While most members did not feel that the organization of the board should be changed, D'Eschambault called for increased francophone representation, including an additional Maritime member to represent Acadian interests. But following "a free and frank discussion," the annual meeting unanimously agreed that its composition should remain as it was.[40] The board did not see the merit of additional members from Québec and Ontario nor from the Canadian Historical Association. Neither did the board see the utility of having a trained historian as its secretary since architectural and engineering expertise was just as beneficial in dealing with problems of conservation.

Despite the impact of the Massey Commission report, the work of the board did not change dramatically overnight. In the postwar years there was little opportunity to expand its work, and few positive moves could be made. And the board continued to be fascinated by traditional topics. In 1952, when asked by the minister to define the criterion "national interest," the board's chairman, Fred Landon, provided a list of categories reminiscent of the board's first attempt at a definition in 1919. Prominent categories were still discovery and exploration, French and English settlement, and the loyalist defence of Upper Canada.[41] Moreover, the board had not come to terms with recognizing historic architecture. Commemoration was still its bailiwick, and so it remained confined to a limited sphere of activity.

The entire program remained confined to habitual approaches. Although the parks branch's involvement in preservation increased dramatically after the war, it was still limited because it had to choose from designated national historic sites, and buildings on this list tended to be military structures. There were other reasons for undertaking the preservation of forts: they were physically imposing, and attracted widespread public attention. There was little reason to stray from the tried and true, and so the first major heritage developments of the 1950s were the restoration of the Halifax Citadel and Lower Fort Garry.

CHAPTER SEVEN

The Politics of Historic Sites in the 1950s and 1960s

The expansion of government activity and the appearance of nationalist sentiment in the 1950s combined to give federal cultural agencies a prominence that they had not previously had. Long-established institutions such as the Public Archives, the National Library, and the National Museum were able to expand their programs, while newer agencies such as the Canadian Broadcasting Corporation and the National Film Board received careful nurturing. In this climate the historic sites program received attention from on high like it had never before. Not just isolated projects, but its larger organization, direction, and goals were carefully examined. This brought a conscious effort to break from past habits, and involved both the Historic Sites and Monuments Board and the National Parks Branch in a more effective policy of heritage preservation. In short order the historic sites program received new legislation, money, and policy; staples it had been without for much of the previous twenty-five years.

In 1953 the Historic Sites and Monuments Act was passed which gave a legislative base to the program. Amended in 1955, the act allowed board members to be appointed from each of the provinces, the Public Archives, and the department as well as one extra member for Québec and Ontario. It also provided for the designation of buildings as national historic sites on the basis of architectural significance, thereby opening a new avenue of program activity. Meanwhile, the department enlarged traditional areas of development, and the historic parks and sites agency became stronger as the department embarked on large scale developments at historic sites across the country.

This activity affected the politics of historic sites as old and new participants sought to define the objectives of a rapidly expanding program. During the 1950s the politics of historic sites became increasingly complicated as old players such as the branch and the board became larger and more active, while senior management of the department emerged as a new participant. Outside the administration there was renewed interest in historic sites as organizations as diverse as the Canadian Historical Association and the Royal Architectural Institute of Canada, along with local pressure groups, presented their particular views on national historic sites. Provincial heritage programs became more firmly entrenched in this period, inserting another layer of official history into the national fabric.

The politics of historic sites became more pronounced as the organizational tangle that had always characterized the heritage program was strained to its limits. Organizational difficulties, such as the status of the historic sites bureau within the larger national parks branch and the relationship of the advisory board to the program, became critical as the department undertook major initiatives in the area of historic conservation and interpretation. A chronic problem facing the historic sites agency was that its administration lay buried in a larger national parks service which, in turn, was part of a polymorphous department, and so had to compete for the minister's attention with other large programs. This handicap was particularly evident at the beginning of the decade. In 1951 the National Parks Branch was one of four large units of the Department of Resources and Development which also included Engineering and Water Services, Northern Administration, and Lands and Forestry. The minister was also responsible for the Government Travel Bureau, Central Mortgage and Housing, and the National Film Board. Although the department was restructured in 1954 when it became the Department of Northern Affairs and Natural Resources and the range of its responsibilities was reduced, parks was never a departmental priority. The parks branch itself was not a homogeneous organization. It consisted of three loosely related divisions: National Parks and Historic Sites, the Canadian Wildlife Service, and the National Museums of Canada. Neither was the administration more focused at the divisional level. The superintendent of historic sites, C.G. Childe, reported to the division chief along with the national park superintendents, the head of engineering and architecture, and the head of forest protection. Later the head of historic sites was made a chief reporting to the director of national parks, but the heritage program still remained a minor part of the organization.

The lowly status of the historic sites bureau was to cause difficulties as the program expanded in the 1950s. Initially, small scale development at national historic parks was left to honorary curators acting under the supervision of the superintendent of historic sites. Then, as development work grew in scale, much of the responsibility for historic park development was assumed by the professionals of the engineering services division. Roads, washrooms, even restoration, came under the direction of the branch's main operational arm. While the historic sites bureau could have directed the activities of engineering services by means of management plans, it lacked the capacity and so abandoned the field of physical development to the engineeers. The bureau concentrated instead on interior development and created a museum section to design interpretive displays using period furnishings. The Historic Sites and Monuments Board, meanwhile, was even less involved than before in these projects. Increasingly, initiatives came from the minister or senior management to be carried forward by the operational arm of the branch, leaving the advisory board to operate on the periphery.

This was an unfortunate situation because the question of national significance, critical for the selection and interpretation of sites, still remained the board's concern. Yet with the bureaucracy and the advisory board functioning on different planes, there was the danger of misunderstanding between the two halves of the program. And, with the expanded activity of the 1950s, national significance became an even more critical problem than before. Just what did this mean? The Massey Commission had seemed confused on the issue, assuming the existence of a unifying national heritage yet criticizing the fact that historic sites were unevenly distributed across the country. The board had struggled with this shibboleth for years without really resolving the question, although it seemed to recommend sites that were associated with important events in regional contexts. The branch, on the other hand, had ignored the problem. In developing sites as national historic parks it had been largely influenced by political or practical considerations. The decisions to develop former British forts in the Maritimes and Québec, for instance, were often taken because they were already owned by the government, they lent themselves to park development, and there was considerable local pressure to have the branch take them over. Because the board had confined itself to the commemorative side of the program, it had not been very involved in the development of historic parks except through the occasional personal intervention of its members. Thus, what criteria the board had managed to muster for defining national significance had been lost in the development of national historic parks.

As preservation became more of a priority during the 1950s, the question of national significance became more of an issue. With the branch embarked in an ever-widening sphere of historic site development, it sought more rational guidelines for long-term planning, especially as the program had been criticized by the Massey Commission for commemorating too few historical themes. Selection criteria especially emerged as a problem in the area of architectural conservation. Although the program became committed to saving architectural heritage, it lacked criteria to sort the exceptional from the commonplace.

THE BRANCH

The recommendation of the Massey Commission that the historic parks and sites agency hire a professional historian was acceded to with the appointment of C.G. Childe's replacement upon the latter's retirement as head of the historic sites bureau in 1954. A.J.H. Richardson was an historian who had been chief of the map division at the Public Archives of Canada and who had a special interest in architectural history. At the end of 1955 the section was made into a division, and Richardson designated a chief reporting directly to the branch director. By the following year two more clerks had been appointed as well as an assistant chief (see appendix 3).

Richardson was supposed to be heavily involved in the development of policy and he was repeatedly asked by both senior management and the board to prepare long-term plans for various aspects of the program. But, like his predecessor, he was overwhelmed just keeping up with the day to day work of the program. As chief he was expected to carry out the mundane duties of a headquarters office: attend meetings, answer queries, supervise staff. As well, he was expected to oversee the work carried on at the widely spread network of historic parks and sites. He advised the local caretaker at Louisbourg on a seasonal program of excavating some of the old walls. He instructed the honorary superintendent of Fort Anne not to use white painted rocks along the border of the entrance road. He nagged recalcitrant caretakers to clean up neglected sites and submit annual reports and advised on the development of a number of museum collections. In addition, he was the principal adviser to the branch on historical matters. Accordingly, he searched the archives for documents to aid in planning the restoration of the Halifax Citadel and buildings in Quebec City. As secretary of the Historic Sites and Monuments Board, he was expected to arrange the annual meetings of the board, correspond with members, and act as a liaison with the department. He was overwhelmed with issues as he was pulled back and forth across the country tending to one emergency after another. It must have been with some relief, then, when he left this position in 1959 to become head of research and passed the reins of this administrative nightmare to J.D. Herbert.

The immediate dilemma facing both Richardson and Herbert was that in order to take on new responsibilities they had to acquire more staff, yet until a new responsibility was formally recognized, they were not going to receive the staff. The problem was exacerbated by the fact that they had to compete for manpower with the entire parks branch. Although the historic sites division was supposed to be on a par with national parks, it was in many ways the poor stepsister. The sympathy of the branch director was critical to solving this dilemma yet indications are that the Historic Sites agency was not his priority. Despite the aid of senior management, the historic sites division had difficulty coping with this blockage, and new positions were created very slowly. In 1958 O.T. Fuller joined the staff as a research assistant, and the following year he and Richardson formed a tiny research unit. By 1962 the research staff consisted of a research director, three historical research officers (one of whom acted as a librarian), and a junior archaeologist[1] (see appendix 4). Shortly after this a museum section was established which was responsible for both the interpretation and conservation of artifacts. Separate units dealt with administration and work pertaining to the board, so that by the mid-1960s the historic sites division was able to cope with long-term planning as well as administer what was already in the system.

But by the mid-1960s it was too late to affect key decisions concerning the form of the program. The 1950s and early 1960s represented a period of intense activity with a number of important heritage properties being developed by the branch. Because the historic sites division was so overwhelmed attending to established sites, it could give little attention to new developments which, by the late 1950s, were largely under the direction of the engineering services division. This division was extremely well organized and operated both out of headquarters and from regional offices. It was the operational arm of the branch, and run with great efficiency by Chief Engineer Gordon Scott. The chief engineer did not take orders from historic sites division, only from the director, which placed the development of national historic parks more within the national parks ambit. Richardson and Herbert were thus placed in the position of consultants whose views regarding historic park development may or may not be accepted by the branch director or the assistant deputy minister. So, in developing policy regarding both the preservation of historic buildings and the development of national historic parks, the historic sites division was at a serious disadvantage.

While the organization of the branch scarcely altered during the 1950s, leaving the historic sites unit largely impotent to formulate policy, changes in departmental leadership helped give the heritage agency a larger profile. The move to more active direction from above reflected a general trend in government that gathered increasing strength from the early 1950s to the present. During the 1950s the government, both under St Laurent and Diefenbaker, was more liable to become involved in large comprehensive activities than at any time before the war. A national housing act, national highway, airlines and communications regulations were all designed to shape the nation along planned routes of development. Regional economic change, immigration, and urban development brought about cultural dislocation, fostering fears of social breakdown, so that by mid-decade the government was receptive to the idea of large cultural programs. This new orientation led to the development of a more complex organization at the top. Large policies were worked out in cabinet committees with inputs from a number of line departments as well as central agencies such as the Department of Finance and Treasury Board. Departmental policies were developed by the deputy minister and assistants drawing on support from the Administrative Services Branch. Structurally, then, it was usual for management above the branch level to develop broad lines of policy. By the 1960s this process became more complex as branches such as national parks incorporated planning units into their own administrative division, but in the 1950s the power resided with senior management.

During this time it happened that senior management was particularly dynamic. In 1951 the minister had been Robert Winters, and the deputy minister was Gen. H.A. Young. The former introduced the bill that finally placed the Historic Sites and Monuments Board on a legislative footing, and Young supervised the acquisition of important historic properties from the Department of National Defence, principally the Halifax Citadel. The department received even more vigorous leadership following its reorganization as the Department of Northern Affairs and National Resources late in 1953. Northern development became a government priority and received special attention from Prime Minister St Laurent.[2] Jean Lesage was appointed minister and, as Young had followed Winters to Public Works, two young men were made senior mandarins. R.G. Robertson had been on the secretariat at the Privy Council Office before coming to the department as deputy minister and Maurice Lamontagne had taught economic development at Laval before joining Robertson as assistant deputy minister. These men set out to master the whole department, not just one or two key agencies, and soon problems of the tiny historic sites operation came to the minister's attention. It is testament to the efficiency of this senior echelon that the minister could introduce a bill in 1954 that would amend the Historic Sites and Monuments Act and bring about profound changes in the orientation of the program.

The historic sites program received more direction from senior management following the appointment in 1955 of Ernest A. Côté who replaced Maurice Lamontagne as assistant deputy minister with special responsibility for historic parks. More than any other individual in this period he placed his personal stamp on the work of his bureau. Few projects were approved until he had visited the site and satisfied himself that it was viable. Rather like Harkin before him, he seems to have had a guiding philosophy and clear ideas about what was or was not appropriate. He was one of the few people at this time who had any systematic idea of the historic sites program and as he remained assistant deputy minister until 1963, he had the opportunity to implement his ideas. He was older than Robertson, still only forty-two years old, but possessed both confidence and energy. Trained as a lawyer, he had risen to colonel during the war and then served with the Department of External Affairs. He had been brought into the new department primarily because of his knowledge of American affairs, and his principal task was to help negotiate the Columbia River Treaty.[3] But his talents made him an obvious choice to bring life to a moribund program. The limitation of his position was structural. Operations were guided by branch officers, and spending estimates compiled by division chiefs. Although he had access to the minister, he could not radically alter operations that were already in place. New money had to be fought for by the minister in cabinet committees or by the deputy minister with Treasury Board. More than anyone else he would have felt the constraints of the Byzantine organization that the bureaucracy was in the process of becoming. Nonetheless, he was in a powerful position to effect change.

Organizational factors meant that in the 1950s policy initiatives were usually generated at the senior management level. While there is little documentary evidence of a coherent policy existing at the top, there are many inferences suggesting the direction that the minister and his assistants wished to go. This nascent policy took two forms: the development of one large historic park project in each province or region, and a program of architectural conservation involving the best heritage buildings in each province preserved through cost-sharing agreements. A related aspect of this policy was the establishment of a network of regional museums. Although this last idea had been dropped by the 1960s, it influenced to a small degree historic site development in the 1950s. The Alexander Graham Bell Museum at Baddeck, Nova Scotia, built in 1954–5, was the most visible manifestation of this impulse,[4] although other regional museums existed at historic parks such as the Halifax Citadel.

By the mid-1950s, then, it was apparent that, in addition to established historic parks, there was interest in initiating a major heritage development in each of the regions, perhaps even in each of the provinces. In St John's, Newfoundland, Signal Hill had been acquired for a national historic park when that province entered Confederation. In Nova Scotia, the Halifax Citadel was obtained from the Department of National Defence and a major program of restoration commenced. The Acadian site of Grand Pré, Nova Scotia, was acquired from the CNR in 1955. On the site was a fancifully reconstructed church where Acadian settlers gathered before being exiled to Louisiana. Associated with the folk hero Evangeline, it was a popular tourist attraction. In Québec City the department acquired considerable real estate from the Department of National Defence and Canadian Arsenals and undertook the restoration of the old city walls and Dufferin Terrace. In Ontario the department obtained Woodside, the boyhood home of William Lyon Mackenzie King in Kitchener, and completed restoration that had been begun by a private organization. The department also seriously considered historic military property in Kingston, although this scheme never materialized. In Manitoba the department acquired control of Lower Fort Garry, although development there did not commence until the 1960s. In Saskatchewan the old North-West Mounted Police post at Fort Battleford was transferred from the province to the department as a ready-made historic park. Further east in Saskatchewan, the old rectory at Batoche was acquired for a museum building. Nothing transpired in Alberta, although the department negotiated with the province to obtain the site of Old Woman's Buffalo Jump as a national historic park. In British Columbia the established historic park at Fort Langley was enlarged, and reconstruction and development was undertaken to coincide with the province's centennial in 1958. Subsequently, the department looked into the possibility of acquiring a military site on Vancouver Island, Fort Rodd Hill, for development as an historic park. The department also explored the possibility of initiating projects in the North and by 1958 was actively investigating Dawson City. These projects shared a number of characteristics: land was acquired *gratis* from other federal departments or provincial governments, the Historic Sites and Monuments Board was not involved in their development, yet local advisory committees or individuals were usually consulted. Finally, the engineering services division played a major role, overshadowing that of the historic sites division.

Of the major projects of the 1950s, the Halifax Citadel and the Québec fortifications stand out as substantial engineering undertakings. Initial estimates for the restoration work at Halifax had been prepared by army engineers for the Massey Commission, and this project became a priority for the branch following the acquisition of the citadel by the department.[5] Initially, the historic sites division had little to do with the work being carried on this site and was more of a consultant than an active participant. Responsibility for supervising the rebuilding of the walls, restoring principal buildings, and stabilizing the remainder was left to the engineering services division. This was the most expensive heritage project of the 1950s, and it had a budget of about $100,000 a year. In 1960 the annual appropriation was increased to $150,000. The engineers were assisted by the chief of historic sites who located old plans in the archives and provided similar technical details. Inside, the interpretation of the site and the organization of the museum displays was left to an honorary curator who was tied as much to a local advisory committee as he was to the branch. A maritime and an army museum were operated by this committee in portions of the old casemates, and the Imperial Order Daughter's of the Empire operated a tearoom in the old magazine.

Côté had endeavoured without success to give the initiative for planning the development of the Halifax Citadel to the historic sites division. In 1956 he sent a memo to the director of national parks in which he said: "I believe that a master plan should be prepared as to the future restoration of the Citadel. There seems to be no such plan in existence. The Historic Sites Division in collaboration with Mr. Ward [the resident engineer] and Major Borrett [the Honorary Curator] ought to propose a plan for restoration over the next few years which should be evaluated as to the monetary costs of various 'slices' and presented to us for approval."[6] But no such plan was forthcoming, and as late as 1961 the chief engineer pointed out to Herbert that, "the only guidance I have received is the approval or rejection of items of work we submit to the Historic Sites Division annually."[7] This issue was fought out during the 1960s with Herbert's staff arguing that archaeological and historical research should precede development, while the engineers maintained superiority in the field. Finally, in the mid-1960s work at the Halifax Citadel was placed on hold while a development plan was drawn up. By this time, however, responsibility for the project had passed to the regional office of the National Parks Branch.

By 1960 the chief of the historic sites division, Herbert, was concentrating his efforts on interior development at major historic sites, leaving exterior work to the engineers. He was in regular contact with museum specialists across the country and developed a focused policy for interpreting historic parks. To implement this policy he formed a core of specialists within his division who concentrated on interior development. The principle behind Herbert's policy in this area was a concept he described as a living museum. He explained such a development as "a house or fort exactly as it was at a given period. An extension of this idea includes 'animation' which might mean having people in period costume doing the jobs and carrying out the tasks which were carried out by the residents of the place at a given period. Our plans for Halifax include the use of attendants dressed as soldiers and artisans working at tasks as they would have appeared about 1860."[8] Subsequently, Herbert dispatched a member of his staff to search the Maritimes for suitable examples of period furniture that could be obtained to reconstruct period rooms at the Halifax Citadel. This approach was in marked contrast to the earlier concept of regional history museums favoured at national historic parks. Living history museums were to concentrate on the interpretation of the site rather than be eclectic displays of artifacts gathered uncritically from the vicinity.

The heritage development at Québec City was even more of an engineering concern. The walls were acquired by the department in 1951, and the branch entered into a cost-sharing agreement with the city and the CPR to rebuild Dufferin Terrace and La Promenade des Gouverneurs, the pedestrian walks linking the Château Frontenac with the citadel and the Plains of Abraham. A resident engineer was placed in charge of this work which continued for the rest of the decade. As at Halifax, the historic sites division acted as a consultant to the project and provided historical data for the engineers. The involvement of the branch increased with the acquisition of Cartier-Brébeuf Park from the city about 1958. This was exclusively an engineering concern as the park had no historic resources. The branch was so involved in the city that by 1958 it had effectively supplanted the National Battlefields Commission as the federal authority in charge of heritage property. Only local pressure kept the body alive, and its chairman was required to submit monthly reports to the director of national parks.[9]

As at the Halifax Citadel, the historic sites division had a slightly stronger voice over questions of development in Québec City. In 1958 the department was informed that Canadian Arsenals was about to declare surplus a complex of buildings known as the Artillery Barracks. The barracks was a complex of buildings inside the walls of the old city consisting of structures from the French, British, and Canadian eras of government. Its historical associations were therefore genuine but vague. The initial investigation was left to Richardson and his assistant Fuller who, in consultation with the local member of the Historic Sites and Monuments Board, argued for the historical significance of part of the complex and drew up a plan for its development as an historic site. The decision on how to proceed with this property was left to the assistant deputy minister responsible for historic sites, Côté, who visited the site in 1959.[10] While sympathetic with the proposal to take over the complex, he was cautious of undertaking responsibility for such a huge development. His answer was to have the department accept the transfer of the property from Canadian Arsenals, and then have the Department of National Defence occupy the buildings as a tenant.

The historic sites division had more control over the modest improvements being carried out at established historic parks such as the Fortress of Louisbourg. From 1952 until 1958 the caretaker at Louisbourg spent about five thousand dollars a year excavating some of the old walls of the hospital and casemates and partly rebuilding them to heights of between two and five feet. Regular reports were submitted to the historic sites bureau which approved the work and authorized expenditure of the year's budget. This arrangement did not suit headquarters management who began to badger the historic sites office for a development plan. In 1955, for example, the chief of national parks sent a memorandum to the branch director stating:

> I have been concerned for some time about the work which is to be done at the Fortress of Louisbourg this summer and have spoken about it to Mr. Richardson on more than one occasion. His trip tomorrow to St. John's is an opportune time to go into our programmme of work at Louisbourg quite thoroughly and I have arranged for him, in company with Mr. Scott [Chief Engineer] and Mr. Gurney [Area Engineer] to spend the best part of a day there for the purpose of discussing and deciding on a comprehensive program.[11]

Côté reiterated the need for a long-term plan the following year. Partly this concern reflected the desire to have a firm policy set down in order to resist local pressure to reconstruct the fortress. The department wished to spend more money there, but within strict limits. Richardson attempted to accommodate this goal by setting out the objectives of the development, which followed the lines already established by the caretaker, and doubling the annual budget to ten thousand dollars.[12] But senior management wished for a still more elaborate scheme, and in 1959 Côté presented a plan to the minister calling for "a basic program of excavating the more important buildings, restoring the original pattern of streets, identifying by illustrative signs the essential features of the town and its fortifications, and extending and improving the museum display."[13] The cost of the basic program was placed at thirty thousand dollars annually for fifteen years. If the minister wished to spend more, Côté suggested incorporating a partial reconstruction into the plan. So, while the work at Louisbourg remained routine, it was left to the historic sites division; major projects were liable to be taken over by senior management and the branch.

THE BOARD

Concern from the top also affected the fortunes of the Historic Sites and Monuments Board in the 1950s. Before the war the board had largely been ignored by the minister and was left to develop a fairly independent role. It was even assumed by the Massey Commission, for example, that the board alone was responsible for heritage policy. Following the war, the board's role as an advisory body to the minister was more carefully defined, reflecting, perhaps, a greater tendency on the part of the minister to initiate policy, and in 1953 it was given a legislative basis with the passage of the Historic Sites and Monuments Act. The act made it clear that the board lacked any executive power and served the minister who could either accept or reject its considerations. This approach continued under Jean Lesage's administration, and the board was given clear instruction about what was expected of it. It was discouraged from making too many designations, and the minister actually rejected recommendations the board had made concerning the commemoration of historic names. The tendency of the minister to control appointments to the board increased during the 1950s. While in 1950 Webster's successor as the New Brunswick representative had been named by his widow, nominations came increasingly from the party machine or the historic sites division.

If the board's power was circumscribed, it gained from greater contact with the minister. As he sought ways to increase the department's profile in the area of heritage, the minister relied on the board to guide him in broad areas of policy. Thus in 1954 Deputy Minister Robertson solicited the board's views on architectural preservation, and in 1957 the minister, Alvin Hamilton, asked for ideas in planning for the nation's centennial celebration. Using the board in this way, the minister had a stake in keeping at least a core of its members as a credible reflection of the heritage community. So while there were always a few weaker members who reflected more political pull than reputation, the board on the whole was stronger during the 1950s than at any time in the past.

Following the death of Dr Webster in 1950, the chairmanship passed to Fred Landon of Ontario. With almost twenty years experience on the board, Landon was well situated to assess its strengths and weaknesses. Although he did not have the forceful personality of Webster, he possessed considerable tact and political sense, qualities that were well expressed in his co-ordination of the board's response to the Massey report. While recognizing the need for reform, he was cognizant of the fallacies of the report's criticism. Landon was also an able representative of Ontario, particularly the western region, and remained a leading figure in the Ontario Historical Society. Neither Newfoundland nor Prince Edward Island had particularly strong representatives in this period and none had connections with heritage groups. Webster's replacement from New Brunswick was A.G. Bailey who, as dean of Arts and professor of history at the University of New Brunswick, brought academic prestige and outlook to the board. Bailey's interest was cultural and native history and he was sympathetic to the idea of architectural conservation. Unfortunately, he did not bring Webster's dedication to the board's work and he was frequently absent from meetings. D.C. Harvey continued to represent Nova Scotia until 1955 when he was succeeded by the very able Bruce Fergusson who also followed him as provincial archivist. Judge Surveyer remained on the board long after age and infirmity had silenced his participation in discussions, and Québec continued to be weakly represented until mid-decade. Reverend D'Eschambault, on the other hand, was a quietly effective representative of Manitoba and became the virtual spokesman for the francophone viewpoint. Campbell Innes, a former school teacher and curator of the Fort Battleford Museum, was appointed to represent Saskatchewan in 1951. Professors M.H. Long and Walter Sage continued to represent Alberta and British Columbia respectively. The scholarly credentials of the board were further bolstered by the presence of W. Kaye Lamb, the dominion archivist, who served from 1949 until his retirement in 1967. Lamb had a PHD in history and was well connected both in government and academic circles. W.D. Cromarty lingered on the board after his retirement from the department until 1953. He was replaced as the department's representative by F.J. Alcock, director of the National Museum. The presence of these last two members was largely honorary, however, and neither was particularly active.

The amended 1955 Historic Sites and Monuments Act which allowed for the appointment of an additional member for both Ontario and Québec immediately served to bolster the French-Canadian outlook on the board. The youthful Edouard Fiset, who joined Surveyer on the board in 1955, was an apt choice for other reasons than vigour. As an architectural planner who had worked with Jacques Grèber in planning the National Capital Region and was presently active in Québec City, he provided a useful professional and regional perspective. Unfortunately, he was dropped in 1960 and replaced by the anglophone publisher of the Québec *Chronicle and Gazette*, Maj. G.C. Dunn. Surveyer, meanwhile, had been re-placed in 1956 by Jules Bazin, a Montréal curator, who was also involved in the provincial survey of historic buildings and was later appointed to the Canada Council. He was succeeded in 1961 by Laval historian Marcel Trudel. Major Dunn was replaced in 1963 by Montréal archivist Jean-Jacques Lefebvre, who was well connected to local heritage groups. Harry Walker, an Ottawa journalist and co-author of *Carleton Saga*, joined Fred Landon as the second Ontario representative.

In 1958 both Walker, who had not been very active, and Landon, who had suffered a stroke, retired to be replaced by a brace of distinguished historians – A.R.M. Lower from Queen's and Donald

Creighton from the University of Toronto. Although busy with other duties, Creighton lent both prestige and wisdom to the board's deliberations. Lower did not participate to the same extent and resigned in 1961. He was replaced by J.J. Talman from the University of Western Ontario who revived a long tradition of strong representatives from the London area. While adding prestige to the board, both Creighton and Lower strengthened the predilection of the board to be a custodian of culture. Both infused their writings with strong nationalist and moral messages, and Lower especially manifested a strongly antimodern bias. Creighton lent his writing talents to produce an official guide to national historic sites. Just the title, *Heroic Beginnings*, is evocative of the Cruikshank era. But despite old-fashioned attitudes, the Ontario historians introduced newer concerns about preserving the built heritage. Lower, for example, was president of a group which sought to preserve homes of fine old Ontario families in the Bay of Quinte area.[14] Landon was replaced as chairman by Reverend d'Eschambault and, upon the latter's death in 1960, Bruce Fergusson was appointed to the chair.

Despite the presence of nationalists like Creighton, the structure of the board confirmed an already established tendency to regard national significance from regional perspectives. This structure, as a collection of provincial constituencies rather than a supraregional body, was confirmed in the legislation which deliberately ignored recommendations of the Massey report and the Canadian Historical Association that the board include more cross-regional representation. The specific recommendation, that the Canadian Historical Association be given the power to nominate two members to the board, also was rejected both by the minister and the board. Although the association lobbied hard to have this principle included in the new act and in the amendment, Landon and the other members counselled against it. The chairman advised Lamontagne, the assistant deputy minister in 1954, for example, that: "I feel very strongly, as do all other members of the Board, that a provincial basis is the only proper one for the Board."[15]

A provincial basis for the board implied provincial perspectives in the selection of sites. This meshed well with the department's desire to apportion sites uniformly across the country but it discouraged attempts to plan sites systematically according to established themes. When Richardson introduced the thematic approach to site selection, it was endorsed by the board as a good idea,[16] but when it came to implement this approach, the board found few common themes. Most themes were rooted in regional development such as British Columbia mining, and the North generally.

So, although the board remained interested in this approach and formed thematic studies and criteria committees, it was wary of universal standards. As before, the board remained deliberately ambiguous on the subject of national significance and seemed willing to accept a great deal of subjectivity. This message filtered through to senior management and, following another recommendation that the department adopt a centralized thematic approach to site selection in 1966, the director of national parks reminded the assistant deputy minister that "[b]oth Dr. Creighton and Dr. Talman of the Historic Sites and Monuments Board have been most insistent that the Department 'stay away from official history.'"[17]

PROVINCIAL HERITAGE AGENCIES

There was one other development affecting the politics of historic sites in the 1950s and 1960s: the rise of provincial heritage agencies. In the early 1950s provincial agencies responsible for historic sites were much as they had been before the war – fairly primitive. If they existed at all they tended to work out of the provincial archives and had minimal operational capacity. An exception was Québec where the government instituted inventories of historic sites. But even this program operated out of the provincial archives in the 1950s. The other provincial agencies mostly supported the work of local heritage societies in acquiring the occasional old building and operating it as a museum.

This situation began to change in the late 1950s. First of all, the provinces began to sponsor large heritage projects of their own. We have seen how in the 1930s the Ontario Niagara Parks Commission undertook the reconstruction of historic buildings in its domain. Ontario became even more involved in this kind of activity in the latter 1950s. In 1954 the proposed St Lawrence Seaway threatened to obliterate an area along the river front that was rich in historic resources. In response to this heritage crisis the province established the St Lawrence Development Commission which, among other things, undertook the removal of representative old buildings from the area to a new site above the flood line. In the ensuing years these buildings were restored and furnished to period condition and the whole complex interpreted as a prototypical pioneer settlement called Upper Canada Village.[18] This kind of outdoor museum consisting of buildings moved onto the site and interpreted by costumed guides, is related to the European folk museum which originated in Sweden in the late nineteenth century. Upper Canada Village was the first development of this kind in

Canada. Another ambitious project undertaken by the Ontario heritage agency was the reconstruction of Ste Marie I at Midland during the 1960s. Although long designated a national historic site and proposed for acquisition by James Coyne, ownership passed to the province which undertook to replicate the vanished seventeenth-century Huron-Jesuit settlement. The province of British Columbia undertook the restoration of part of Barkerville, a pioneer village *in situ,* to celebrate its centennial in 1958. Barkerville had been the largest settlement on the mainland soon after the gold rush started but just as quickly it been left a ghost town. The province refurbished a core of this town and developed it as a major tourist attraction. A different kind of initiative was taken by the province of Québec in 1956 when it introduced the concept of a designated heritage monument. Based on French legislation, designated monuments were privately owned buildings recognized by the state and therefore protected by heritage building codes.[19]

Following initiatives like these, the provincial agencies began to change organizationally, and, in the process, instituted more comprehensive programs of development. The first province to reorganize bureaucratically along these lines was Québec which established a ministry of culture in 1961. Other provinces followed suit in the 1960s and 1970s, although usually the cultural agency which included historic sites, archives, and museums was combined with programs of tourist development and recreation. The historic sites part of these agencies closely resembled the federal heritage program: historic sites were officially recognized and commemorated, sometimes with the assistance of an advisory board, and historic parks were established as recreational and tourist developments. At first these sites were regarded as junior to the national sites, and there was ample opportunity for a kind of co-operative federalism in this field. Indeed, in 1961 federal and provincial deputy ministers responsible for heritage programs established the Canadian Conference on Historic Resources which met annually thereafter. But as provincial identities strengthened during the 1960s some provinces took the position that they had exclusive jurisdiction over historic resources. Certainly this was the position of the Québec government which presented its case at the First Ministers' Conference in 1967.[20] This attitude away from co-operative federalism meant that the provincial agencies tended to become rivals instead of junior partners of the federal government. This would have serious implications for co-operative undertakings in the area of architectural preservation as we shall see in the next chapter.

CHAPTER EIGHT

Conserving the Architectural Landscape, 1954–67

Historic park development only partly assuaged the needs of the heritage community in the postwar era. While there was unanimous support for the preservation of large military establishments like the Halifax Citadel which were important to the city's sense of its past and future tourist development, these projects did little to preserve historic architecture that was disappearing at an alarming rate across the country. The postwar boom inspired the redevelopment of Canadian city centres such as Halifax, Québec, Montréal, and Vancouver. With this transformation, stately homes were demolished to make way for commercial development, and historic districts were threatened by urban highway systems. A development mentality took hold where anything new was perceived as good, while old buildings were bad, reflecting stagnation.[1] This outlook even affected Ottawa. The old supreme court building, on the western fringe of Parliament Hill, was torn down and the Department of Public Works seriously considered demolishing the west block of the parliament buildings.

Despite this massive impulse to eliminate the built heritage, or perhaps because of it, isolated pockets of resistance gathered strength during the 1950s to oppose the trend. The Architectural Conservancy of Ontario, formed in the 1930s, continued to be an important lobby in the province. Across the country, local groups mobilized to save period houses and historic neighbourhoods. Some of these groups looked to federal and provincial governments for support. And in 1959 the Royal Architectural Institute of Canada formed a preservation committee chaired by Eric Arthur.

This impetus within the heritage movement had influenced the recommendations of the Massey report and subsequently helped convince Lesage and his senior staff to initiate a policy of architectural preservation. What was needed, however, was a completely different kind of activity than that traditionally carried out by the federal agency. Architectural preservation could not be effectively achieved through the placing of historical markers, or the creation of historic parks. This need for a distinctly new activity led senior management to bypass established bureaucratic structures in the National Parks Branch and seek advice directly from the advisory body. And this connection between the minister and the board greatly enhanced the power of the advisory body. In contrast to the historic park side of the program, where the initiative had been largely taken over by the National Parks Branch, the new activity of architectural preservation promised to involve the board to a much greater extent. Ironically, this new activity also promised to give greater influence to the historic sites division which had little power over new historic park development but which was tied to the board through the division chief who acted as its secretary. The preservation of historic buildings, therefore, evolved as a much more independent activity than the development of national historic parks which remained tied to the national park program. Before proceeding to examine this new activity, we should look at the relationship between the minister and his advisory body.

In moving to preserve something of the rapidly disappearing architectural past, the federal government naturally looked to its established heritage program. This, in turn, brought a range of traditional approaches to the problem. In a sense the situation resembled concerns about destruction of the built environment which had surfaced at the beginning of the century and which had partly inspired the formation of a heritage movement. This had led to the establishment of the national historic parks and sites program and had initially moved the National Parks Branch to seek the help of provincial agencies in compiling a list of places deserving preservation and to draft a bill that would have given protection to designated properties. That these initiatives were abortive did not mean that the idea of architectural preservation had died out, and when the federal government sought a policy for the preservation of historic architecture, the program offered up these two old chestnuts from the past, legislation and a national inventory. The Historic Sites and Monuments Act as amended in 1955 allowed for the designation of a national historic site on the basis of architectural merit. Following this, an inventory of historic buildings was begun to aid in the selection of meritorious buildings.

The problem here was that the program only offered one way of preserving property, through its acquisition and development as a national historic park. This was a costly and inefficient way of preserving domestic architecture, and the program would not have made much of an impact across the country by taking this approach. Designating a building as a national historic site would lend the moral weight of official recognition but offered no tangible form of protection. Here the federal government was handicapped in what it could do because under the separation of powers it had little authority over private property. This was a provincial jurisdiction. The ideal solution was for the federal government to engage in some kind of co-operative endeavour with provincial and local organizations to preserve selected buildings across the country. This is what the federal program attempted in the latter 1950s. In doing this, though, it took on a number of practical and theoretical difficulties. A major conceptual problem was site selection; if sites were to involve the co-operation of provincial and local agencies, how could they be selected according to a single national standard?

THE SEARCH FOR A POLICY

This issue of single *versus* multiple standards of site selection affected the question of architectural policy from the beginning. In 1954 senior management indicated to the board that it wished to comply with the Massey report's recommendation that the department preserve distinctive examples of Canadian architecture.[2] It was preparing an amendment to the Historic Sites and Monuments Act that would allow the minister to designate buildings as national historic sites for reasons of architectural significance and wanted the board to prepare a policy on the subject. In the ensuing discussion between the department and the board it became apparent that two things were being sought. On the one hand, the department hinted that it wished a regional allocation. At one point, for example, Côté asked: "If, for example, there was to be one large building selected per Province, what would be the continuing cost of maintaining such a building."[3] On the other hand, by seeking a criteria for selection, the department also sought a consistent standard. The dilemma was put to the board in a letter from Richardson to Landon: "Should the program be directed toward preserving samples from each province or important area, or towards preserving a representative collection of types and styles?"[4]

Presented with a unique opportunity to redirect the heritage program along new lines, the board emerged as a more potent body for the board's inclination parallelled that of senior management in preferring to emphasize preservation over commemoration through the creation of plaques. At a board meeting in June 1957, for example, Landon "reminded the members of the great field of work opening up in the restoration and preservation of buildings. He agreed with the opinion of certain members that tablets are too much of the one kind."[5] Even A.R.M. Lower, who, one would think, would have favoured plaques for their morally didactic purpose, seemed to be in favour of a reorientation toward greater emphasis on preservation. Upon his resignation in 1961 he informed Walter Dinsdale, the minister, that: "I think we would be better advised to concentrate our money on the maintenance of historic sites as a whole. I have in mind such features as battlefields, forts, historic houses of good architectural merit and possibly even whole sections of streets in such cities as Quebec and Kingston."[6] Given this attitude on the board, it is not surprising that it lost no time in accepting the challenge to devise a new policy.

The board worked out its policy on architectural preservation between 1954 and 1956, presenting a detailed report to Lesage, the minister. Subsequently, the board worked on perfecting this policy, individually through its members and through a historic buildings committee established in 1958. The board's policy consisted of four main recommendations: it suggested comprehensive criteria for the selection of architecturally significant buildings; it urged the government to become involved in architectural preservation through cost-sharing agreements with other levels of government and local organizations, it urged the historic sites agency to initiate a national inventory of historic buildings, and it recommended the establishment of a national trust along the lines of the British organization. In formulating this policy the board maintained close ties with senior management and came to favour a decentralized approach to identifying architectural heritage. This was in contrast to the more homogenous stance favoured by the historic sites division.

Initially, it seemed as if the board preferred a uniform standard for identifying architecturally significant buildings rather than a regional allocation. Its policy, sent to Lesage in the spring of 1956, advised that a building to be designated of national architectural significance should exist "in its original form, be venerable by reason of age and tradition, or enriched by its association with events of national historic importance, or remarkable as a type of existing architecture, or a form of building that has influenced the art of architecture in Canada."[7] But it soon appeared that the board's approach to building selection was ambivalent. This ambiguity was particularly apparent in the structure of the building survey which it attempted to initiate in 1955–6. The building survey or inventory was supposed to form the basis for selecting buildings for designation as the best surviving types of buildings. It was, therefore, an important point in the board's building policy submitted to the minister. But while the inventory approach may have seemed to provide an objective or even scientific solution to the problem of selecting buildings for national recognition, in fact it merely concealed and did not resolve a fundamental question of selection. Was this survey to be a national one, or a series of provincial surveys? Was it to include universal styles, or regional building types? The board tried to accommodate both sides of the problem. Its members surveyed their respective provinces while calling for a national inventory. In November 1958 a board committee, "Established to Study the Preservation of Historic Buildings," suggested a list of building categories that could include both national and regional types in a national inventory: seventeenth century, French seventeenth to nineteenth centuries, Acadian, log, sod, pre-Loyalist, Loyalist, Georgian, Victorian, Greek revival, Gothic revival, and twentieth century. At the same time the committee recommended that "each member of the Board ... after consultation or communication with the Secretary of the Board, provincial and local bodies, historical societies, societies of architects or other organizations or individuals, forward a list for his province or region as soon as possible."[8]

The department, too, had different ideas about the basis of the inventory. Côté felt that the inventory should have a provincial basis and perhaps lead to a provincially organized national register of historic buildings with the top few on each provincial list being eligible for federal assistance. In a letter to a provincial colleague written in 1960 he stated that: "When it comes to making an inventory of all historic buildings in a province, this would seem to be primarily a provincial job as few buildings can be classed as of national significance."[9] Côté's approach to an inventory reflected his belief that the department would become involved in architectural preservation on a regional basis.

The historic sites division, on the other hand, was more inclined to view architectural heritage from a single national perspective. This approach is reflected in a query from A.J.H. Richardson to the director of the McGill School of Architecture about a house in Halifax.

The Department is at present working out a set of criteria to apply to early buildings to determine whether they are of national historic importance on historical or architectural grounds, but these have not yet been finalized, and I am not asking for your opinion whether the Round House is of national importance, only for any assessment you would like to make of its significance in Canadian architectural history, whether you think it is an exceptionally good or very characteristic example of the architecture of the period in Canada, and whether it is a characteristic or a unique remaining example of an early pleasure building of this sort.[10]

This approach of the historic sites division coloured its approach to a national inventory. Whereas Côté had envisaged a series of provincially based registers of historic buildings, Richardson was inclined to follow the example of the Historic American Building Survey operating in the United States. The survey had been initiated in the 1930s as part of Roosevelt's New Deal to provide work for unemployed architects and photographers. Teams covered the country looking for interesting examples of architecture which were then recorded by photographs and measured drawings. These records ended up in the Library of Congress. This project was resurrected in the 1950s by the US National Parks Service in co-operation with the American Institute of Architects and the Library of Congress. Preservation, however, was only a secondary aim of this inventory. Its initial purpose was to provide an accurate record of a vanishing past of America. Through the 1960s Richardson strove to launch a similar enterprise through the co-operation of the Royal Architectural Institute of Canada and the Public Archives. Following a series of pilot studies of buildings in historic urban areas carried out in the 1960s, a fully fledged computerized survey – the Canadian Inventory of Historic Buildings – was instituted in 1970. But, like HABS, its American counterpart, the CIHB began and ended with the recording process and had little impact on heritage preservation.

The tension between national and provincial perspectives affected another tenet of the board's policy on architectural preservation, the establishment of a national trust. The board had in mind a national body, along the lines of the British society, but with more government support, that could acquire outstanding examples of Canadian architecture. The Saskatchewan member, Richmond Mayson, was particularly interested in this concept, although it had been suggested by the Historical and Cultural Committee of the Canadian Tourist Association. In 1958 and 1959 Mayson met with a group of influential Montréal businessmen, which included C.J.G. Molson, in an effort to establish the trust. But in the end it was decided "that it would serve to better advantage to have matters organized in the provinces first and then bring all units together as a National Heritage Trust."[11] This is what ultimately transpired, although it took until 1973 to establish an organization along these lines.[12]

The interest in a national trust was tied to another belief shared by both senior management and the board: the department should not undertake permanent responsibility for architecturally significant buildings. This message was spelled out in a letter to Lesage, the minister, from Landon in 1956 in which he remarked that the board members "were unanimous in their view that no building should be acquired that could not be put to some practical use and equally one in mind that there should be the fullest co-operation with other authorities, provincial or municipal, in this work."[13]

The problem with the desire of the board and senior management for a decentralized and co-operative program of architectural conservation was that it depended on the existence of locally organized participation and there was little of this until the latter 1960s. This situation was remarked upon in a letter from Alvin Hamilton, the Minister in 1960, to the author of a magazine article that had been critical of Canadian conservation policy.

> What is clearly needed if, as you have urged in your article, the agencies of governments should move more quickly to protect our historic landmarks, is a greater public awareness of the 'facts of life' in regard to historic preservation: which buildings are important to them and why; to what use can these be put; which levels of government have authority to act in this way, and how; should they have more authority or more funds for this purpose; what determines whether the federal government should set to preserve a certain building, or whether this should be up to the provincial or municipal authorities, should a greater role be played by non-governmental organizations in this work, and do we need more and stronger voluntary associations formed for this purpose.[14]

Without this grass roots awareness and support, comprehensive co-operative programs for architectural conservation would be a long time in coming.

While the department and the board awaited the completion of the inventory and the establishment of a national trust, they participated in a number of initiatives that went a long way toward involving the program in architectural conservation in a meaningful way. These initiatives conformed to precepts already accepted by both senior management and the board: they were considered by the department to be among the top ten architectural examples in the province, they were identified by the board as having national historical interest and recommended for preservation. Most important, all of these initiatives involved cost-sharing agreements where the federal government provided part or even all of the capital costs, but ongoing operation and maintenance of the building became the responsibility of local or provincial organizations. Because overall planning had not been completed, these initiatives tended to occur where there was strong local support with active representation on the board. Consequently, the dozen or so agreements initiated between 1958 and 1967 were clustered in New Brunswick, Québec, Ontario, Manitoba, and British Columbia. Despite active support in Nova Scotia, the government was unable at this time to conclude an agreement to conserve buildings along the Halifax waterfront. The old town clock, however, was included within the boundaries of the Halifax Citadel National Historic Park.

The first of the initiatives to conserve urban architecture involved a building at 17 St Louis Street in Québec City called the Maillou House. The building had been a prominent residence during both the French and English régimes and so was well documented in historical records. It had been constructed in 1736 by a noted Québec builder named Jean-Baptiste Maillou, architect of the Hôpital génerale. The house was subsequently enlarged and, following the conquest, occupied by the deputy postmaster general for the British army establishment in North America and then by the army commissariat headquarters. The building was then acquired by the Canadian military and in the 1950s contained offices of the Canadian Signals Corps.

The preservation of the Maillou House was realized as a result of the coincidence of favourable circumstances. The owner, the Department of National Defence, had declared the property surplus to its needs. The department had been disposing of a great deal of real estate in the city, and the Department of Northern Affairs and National Resources had already acquired surplus military property. Therefore, the historic sites bureau was given first refusal on the property. The local board member, Edouard Fiset, was an architect also active on the Québec Board of Trade which, in an effort to boost tourism, was trying to promote the development of the city's built heritage. Fiset had personal contact with Assistant Deputy Minister Côté and so was well placed to propose informally an agreement where the department would acquire and help restore the property, and the Board of Trade would care for the house through a long-term lease. The house would be used partly for offices and partly as a museum open to the public.

Fiset's scheme was formally proposed by the board in the form of a motion passed in May 1958 recommending "that the Department of Northern Affairs and National Resources accept the transfer of the property from the Department of National Defence, and study the possibility of placing it in the custody of a responsible public body for a period of ten or fifteen years."[15] Hamilton, the minister, agreed with this recommendation and in July he announced that the department would accept transfer of the property from the Department of National Defence and put the lease of the property up for tender. Meanwhile, the research component of the historic sites division, namely, Richardson and his assistant, Fuller, compiled a detailed history of the building. Senior management sought a more definite recommendation from the board which, at its November meeting, passed the following resolution: "Of the buildings now available for protection in the Province of Quebec, the Board considers that the Old Commissariat House at 17 St. Louis Street in Quebec City, is in priority, among the first ten buildings worthy of preservation on grounds of national historical and architectural importance."[16] In December Hamilton announced that an agreement had been reached with the Québec Board of Trade to preserve the building.

The preliminary scheme for developing the property had been prepared by Richardson in consultation with the resident engineer in Québec. This proposed to restore the exterior facade to its appearance in 1831, while partially restoring the interior and furnishing two of the rooms with period antiques. The "guesstimate" of the costs of this work was $135,000. Under the terms of the cost-sharing agreement, the Board of Trade assumed responsibility for all of the restoration. The architect was to be hired by the Board of Trade and given responsibility for preparing the plans. But as $95,000 was to be provided by the department, it remained deeply involved in the project. The resident engineer of the parks branch in the city acted as project manager, and Richardson of the historic sites division was in regular consultation with both the architect and the engineer, providing specifications and details for the period restoration.[17] This, in fact, was the first major restoration work in which the historic sites division was directly involved.[18]

Difficulties arose during 1959. It soon became clear that the $135,000 estimate was far too low, and approval was sought to spend $171,000. Then it was discovered that developers wished to rebuild on some surrounding property in a way that would have diminished the effect of the restoration. The Historic Sites and Monuments Board protested against this planned development and the government took measures to control the adjacent property which was owned by the CNR and the city. The Maillou House was not only the first manifestation of the department's policy on architectural preservation – it was the beginning of massive federal involvement in the preservation of the city's heritage architecture.

A second important agreement for the conservation of a heritage building occurred in Manitoba. The initiative to conserve the old Grey Nuns' convent in St Boniface began as a desire on the part of Mgr D'Eschambault to have the federal government pay for a building for the museum of his St Boniface Historical Society. A board recommendation to this effect was duly made but was rejected by Hamilton.[19] D'Eschambault persisted and received a creative idea from Assistant Deputy Minister Côté which he related to Richardson in September 1957: "I have been advised by Mr. Côté that we should suggest the taking over of an old building of some historical character and this could be adapted for the need. The oldest building actually in use and which has never been really changed, is the Provincial House of the Grey Nuns ... The Sisters arrived here in 1844 and their house was begun two years' later."[20] Subsequently, D'Eschambault had the board recommend the designation of the building as a national historic site at its meeting in November 1958. Then, D'Eschambault attempted to obtain a cost-sharing agreement among the St Boniface Historical Society, the city of St Boniface, the province of Manitoba, and the federal government. But this project was not destined to fall into place as easily as the Maillou House. It proved difficult to implement with so many players and was aggravated by a complicated scheme to relocate the building to an adjacent park.

One initial difficulty was that the historic sites division was decidedly lukewarm about the idea. It was planning a large museum development at Lower Fort Garry and did not relish the idea of another museum so nearby, possibly funded from its budget. In August 1959 J.D. Herbert, who had succeeded Richardson as division chief, sent a memorandum to the branch director, J.R.B. Coleman, expressing his disapproval.

> In the first place, their claim that this is the oldest occupied building in Western Canada is not true. The Factor's House at Fort Garry is at least twelve years older and has been continually occupied. Mr. Richardson tells me he believes there are other buildings in the Red River Valley which are also older. The fact that the building has been a convent of the Grey Nuns ... does not give it national historical significance ... This venture looks like an effort to achieve the previous objective by a backstairs

method. The only proper course to follow, in my opinion, would be to advise the Minister to defer his approval of the Board's recommendations until further research casts more light on the picture.[21]

But the division was unable to sustain this unco-operative position in the face of continued pressure from D'Eschambault. Testimonials supplied on his behalf by the Director of the University of Manitoba school of architecture and Gérard Morriset, curator of the Québec provincial museum, placed the division on the defensive. Herbert then sought independent advice from the director of the Glenbow Museum, Calgary, asking him to comment on the building in the context of the province's ten best. Clifford Wilson's reply seemed to finish the matter. While he acknowledged that the building was not the oldest in the province, he added that "[i]t is beautifully proportioned and I certainly hope it can be preserved."[22]

Having accepted the validity of preserving the building, the division balked at the idea of moving the building. It influenced the department to negotiate for the preservation of the building *in situ*. A letter to D'Eschambault, drafted by Herbert and signed by Hamilton, argued the importance of maintaining the integrity of the site and building.

Together they account for certain intangible elements of feeling and association without which the end result tends to be without life and character. I am trying to formulate a policy for the future in which these roughly hewn ideas will become foundation stones ... I therefore feel that any assistance offered by the Government of Canada for the preservation and restoration of the convent would have to be predicated on the understanding that the building remains on its present location."[23]

If this difficulty could be overcome, Hamilton added, the government was prepared to contribute 40 percent of the total estimated cost of $100,000. With this incentive, the local group strove to arrange for the preservation of the building on the site. Meanwhile, D'Eschambault passed away, leaving the project to become his monument.

The decision to keep the building on the site delayed the agreement still further, and protracted negotiations among the historical society, the church, and various levels of government continued through 1961. Throughout these proceedings Assistant Deputy Minister Côté played an active role. Finally an agreement was reached where the church was to cede the land with the convent, while the city, in co-operation with the federal government, would restore the building, and the historical society would operate a museum on the premises.

Although the federal government was to contribute less than 40 percent of the capital costs, and restoration was theoretically the responsibility of the city, in fact responsibility for planning and development fell on the shoulders of the historic sites division. The general mandate of the division was outlined in a memorandum from Herbert to Richardson in January 1962: "Mr. Côté has committed us, as you will see, to providing specifications for the restoration of the house. 'Specifications' may be translated rather loosely. What he has in mind is that we should tell the City what should be restored in the way of illustrating the historical uses of the building, and, if possible, point these out and give some guidance as to the appearance of these rooms at the height of their functioning."[24]

The problem facing the division in carrying out this task was that it had little formal authority over the project and little operational capacity. Its first task, for instance, was to obtain a more detailed and reliable estimate of the restoration costs. To do this it had engineering services division detail an engineer from its western regional office based in Banff. The result was not encouraging. The project engineer reported to his superior in Banff, who notified the chief engineer in Ottawa, who told the branch director, who informed Herbert that: "In my opinion the present building would be useless as a Museum and the cost of putting the building in suitable shape would be prohibitive, and would likely cost more than a smaller new building."[25] The historic sites division was stunned by this news and hesitated. Côté, on the other hand, was not amused and insisted that the division take action. Fortunately, the division was able to call on a consulting restoration architect, Peter John Stokes, who had advised on other similar projects, to visit the site. His report was not nearly so gloomy as the engineer's, and the department went ahead with the project. Still, events moved slowly: the architect's plans were not completed until November 1964, and in March 1966 another sixty thousand dollars was needed to finish the restoration. By this time the collection of the St Boniface Historical Society was considered to be too meagre and eclectic for a building with such a particular history, and it was decided to place a traveling exhibit, prepared by the historic sites division for the centennial year, in the restored building instead.

By 1967 a number of other cost-sharing agreements, involving various degrees of federal support, had been concluded for the preservation of historic architecture. The buildings included the

Leonard Tilley House in Gagetown, New Brunswick, a loyalist house in Saint John, military barracks in Fredericton, the Black Binney House in Halifax, a lockmaster's house in Montréal, another loyalist house in Williamstown, Ontario, the Gage House in Hamilton, the Mather House in Perth, the Ermatinger House in Sault Ste Marie, and the Craigflower Manor and Emily Carr House in Victoria, British Columbia. As well, the department authorized the occasional grant to allow municipalities to rehabilitate civic structures, and gave $100,000 to the city of Kingston for the restoration of the portico of its city hall. Despite many difficulties, senior management and the board had combined to initiate a reasonably coherent policy of architectural conservation. It was never intended that these initiatives would accomplish all that was needed in preservation but, as Côté informed the minister, Arthur Laing, in 1967, referring to the Halifax waterfront development then being negotiated: "It would seem to me that for the time being the best influence that the Federal Government can exert in this general area of urban and 'historic complex' work is to create catalyst type projects."[26]

Despite these early initiatives, the department was not completely clear about its preservation policy. In a memorandum to the branch written late in 1960, Côté noted:

> What we lack at the present time is a written policy as to the preservation of buildings of *national* historic importance and regarding aid (if any) for the preservation of buildings which are not in this category. My guess is that we could not properly tackle a program for the latter category until we had a clear policy for the former.
>
> We have in at least embryonic form, fairly clearly in mind what should be the policy for the preservation of buildings, but we have not got it spelled out in a manner which we would wish to transmit, say, to Treasury Board or indeed to the general public.[27]

He was concerned that funding for buildings remained on an *ad hoc* basis because there was no regular budget for ongoing preservation. But without a written policy, it was easy to lose sight of recently evolved objectives. Thus in the same month that Côté's memo appeared, a policy statement regarding building preservation said: "The policy pursued by the Government of Canada regarding the preservation of *buildings* is first to determine which buildings are of *national significance* ... A selected number of buildings are then preserved as National Historic Parks."[28] As before, stated policy was drifting away from practice.

One difficulty with this decentralized approach where the department worked through cost-sharing agreements was that the heritage program maintained a low profile. Federal involvement in these developments was often forgotten after the important start-up work. With the advent in the 1960s of huge historic sites projects such as the reconstruction of the Fortress of Louisbourg and the restoration of Dawson, cost-sharing agreements became overshadowed by national historic parks developed solely within the organization. Nonetheless, they represent a creative and innovative approach to heritage preservation.

Another difficulty was that in undertaking the preservation of privately owned buildings the government could be perceived to be entering the domain of the provinces. Certainly by the 1970s other provinces besides Quebec, notably Alberta and Ontario, had instituted their own programs of architectural preservation. The answer to this problem was federal-provincial agreements and co-operative arrangements along the lines of the American National Register of Historic Buildings where state heritage agencies participated in nominating buildings to a national register, making them eligible for federal funding and placing them under legislated protection. But the competitive atmosphere of the 1970s made federal-provincial agreements of this type impossible to achieve.

Ottawa in part circumvented this last problem by participating in the establishment of Heritage Canada in 1973. The federal government provided a one-time grant of twelve million dollars and technical and administrative support from its national park agency. Heritage Canada did not set out to behave as a monolithic agency but has endeavoured to work through local organizations much like the Historic Landmarks Association many years before. In promoting grass roots projects, Heritage Canada has not instituted national criteria for the selection of heritage buildings but has supported local initiatives to save individual buildings and preserve and develop heritage areas. By the same token it does not dispense funds for these projects and uses its monies to support an ongoing organization.

CHAPTER NINE

The Era of the Big Project, 1960 and Beyond

During the 1950s the historic sites program evolved into a more complex organism as it absorbed more historic park developments and embarked on a new policy of architectural preservation. With this growth the key players tended to become stronger. The minister and his deputies became involved in new initiatives to an unprecedented extent. The historic sites bureau grew in size and competence to better manage the special resources under its care. The Historic Sites and Monuments Board shook off its diffidence and for the first time since its inception addressed the question of preservation. And local groups began to mobilize, to recognize and care for the built heritage. All of this activity strengthened the program as a whole and influenced its directions. By 1960, then, the program had gathered considerable momentum and was picking up speed.

But this growth led to another plateau of activity, one that took the organization in a new direction, and in the process, wrought lasting structural changes. This new era was marked by the big project, the development of large heritage properties completely under the department's control. Until this time historic parks had always been an important part of the program but they had never defined its organization. Most of the historic parks had been in the system for a long time and had undergone only modest or sporadic development. True, there had been a flurry of activity during the late 1930s when Port Royal Habitation had been reconstructed, but this was an exception. During the 1950s, while there was widespread activity at historic parks across the country, development was still modest by today's standards. Sustained initiatives were directed toward commemorations or cost-sharing agreements for architectural preservation. But during the 1960s a whole chain of development was initiated at several historic parks across the country that lasted well into the next decade.

This new phase captured the attention of senior management and the public, reoriented the development of the historic sites administration which evolved to meet this new challenge, and relegated the advisory board to its former small corner of commemoration. The earlier policy of architectural preservation was largely eclipsed during this new era, and the department passed much of the responsibility for this part of the program to Heritage Canada which emerged in the 1970s as a kind of arm's length government agency. Only the Canadian Inventory of Historic Building survived as a vestige of the old direction. It is to the big projects, then, that we must turn our attention.

During the 1960s the historic sites program became dominated by a number of large projects: the Halifax Citadel, the Fortress of Louisbourg, Lower Fort Garry, and Dawson. Initially, as in the 1930s during a similar period of development, the operational side of the National Parks Branch tended to bypass the previous policies of the historic sites division. The Historic Sites and Monuments Board, too, found itself in the role of a bystander in the new program as the branch director developed lines of action to comply with ministerial requests. This tendency was most evident where there was unusual haste to make a substantial showing. Two projects, for example, the reconstruction of the Palace Grand Theatre in Dawson, and the reconstruction of part of the Fortress of Louisbourg, had to be developed extremely quickly to meet ministerial commitments. In these circumstances the director and the engineers were chary of lengthy consultation and eager to begin work. In both instances, too, there was potential for disagreement between the director and the engineers, who viewed the projects more as tourist developments, and the historic sites division and the board, which were more conscious of principles of conservation. Eventually the historic sites division had to be enlarged and associated with these projects as the necessity for careful and systematic planning became evident and as these projects became the flagships of the agency. But, because the projects were initiated by forces outside the agency, they took the program along different paths than the one it had been following alone in the 1950s.

The two big projects, at Dawson in the Yukon and Louisbourg on Cape Breton, were initiated by larger government policy. The suddenness with which they were presented, and the operational problems they posed placed enormous stresses on the historic sites agency. Moreover, they focused on issues about the nature of the heritage program as only an atmosphere of crisis could. Only gradually did the branch and then the historic sites division assume control of the projects and devise their own policies of development.

The first of these projects, the reconstruction of the Palace Grand Theatre at Dawson, emerged from the government's larger interest in northern development. In the mid-1950s Minister Jean Lesage had made this area a priority and asked the board to consider ways in which historic sites could be established in the North.[1] The board looked into the feasibility of establishing regional museums at Whitehorse and Yellowknife[2] and, although nothing further materialized in the Northwest Territories, the board sent its British Columbia member, Walter Sage, to the Yukon and recommended the acquisition of a stern-wheeler to house a local museum.[3] Interest in the North then diminished until after the election of the Diefenbaker government in 1957.

Prime Minister Diefenbaker seemed to regard the North even more positively than St Laurent. Describing Diefenbaker's national vision, Bothwell, Drummond, and English have remarked: "Some of this related to social welfare measures and to the equalization of opportunity and development throughout the dominion. But most related to the North, and it was Diefenbaker's northern vision that captured the national imagination."[4] During a visit in April 1959, the prime minister himself first raised the possibility of developing Dawson City as an historic tourist attraction. Aware of the tourist potential of the Yukon that had opened up with the recent creation of the state of Alaska, he pointed out the attraction of the heritage gold rush for Yukon visitors. He even expressed the opinion that the old auditorium should be preserved as an historic site.[5] Quite suddenly, then, the historic sites agency found itself committed to a building and a line of action without having had the opportunity to reflect and weigh the consequences.

Initiative passed from the prime minister to the minister, Alvin Hamilton, who asked the board for its opinion on Dawson City and the auditorium.[6] Walter Sage was again sent north to report, and at its November 1959 meeting the board advised that Dawson City was of national historic interest and the auditorium should be preserved. Meanwhile, Hamilton engaged in discussions with Tom Patterson, one of the founders of the Stratford Festival, to establish a Gold Rush Festival which would be based in the old buildings at Dawson. In the words of the deputy minister, Gordon Robertson, Patterson "thought that such a festival could be built up, over a few years as a permanent living re-creation of the 1890's that so stirred the nation's development and history."[7] Patterson was engaged as a consultant in October and submitted his report in November 1959. He recommended centering the festival in the auditorium and developing side attractions: "visits of historical sites, gold panning contests, historical film festival, visits to sternwheeler etc."[8] Events, it seemed, were overtaking the ability of the Historic Sites and Monuments Board to respond.

In March 1959, Hamilton announced his decision to support a gold rush festival through the restoration of the Dawson auditorium. A private organization, the Dawson Festival Foundation, was founded to co-ordinate the gala event, and planned to start in the summer of 1962. The department's task was to prepare the auditorium in time for the event, although there was implicit concern for the "side attractions."[9]

J.D. Herbert, chief of the historic sites division, along with the regional engineer visited Dawson in June 1960 to assess the nature of the task. They estimated that approximately $175,000 would be needed to restore the building.[10] A later report from the engineers advised Herbert that he should not count on having the auditorium ready for July 1962.[11] The problems involved in carrying out restoration work in the North were just too great. Matters were further complicated by lack of funding. The department would have to approach Treasury Board for supplementary funds, a process that could take weeks if not months. But Assistant Deputy Minister Côté advised his staff that "we must try to complete for summer of 1962."[12]

At the end of 1960, the branch gave Dawson top priority and rose to meet the challenge laid down by Hamilton. In so doing it bent or ignored established practice. Branch officials met to discuss ways in which funds could be raised to initiate work immediately and proposed borrowing from other projects.[13] In January a contract was let without tender to a Vancouver architectural firm to do a feasibility study. The report was ready at the beginning of February, estimating the cost of reconstruction at about $290,000.[14] The architects addressed the question of reconstruction versus restoration and convinced the branch that it would be cheaper, quicker, and easier to demolish the old building and replace it with a modern replica. A deciding factor in this issue, besides time, was the strict safety code imposed by federal fire regulations. By law, functioning theatres have to have ample stairways and exits and be equipped with a sprinkler system. Modifying the existing structure to accommodate these features would have been difficult and costly. Better to build from scratch, the architect's argued, so that necessary pipes and fittings could be concealed in the framework. The decision to reconstruct was made very quickly in Ottawa without consulting the board and the same Vancouver firm was given another contract to manage the reconstruction of the auditorium on 10 February.[15] Côté was assured that the building would be ready for July 1962, and that "[t]he reproduction will be exactly authentic" save, of course for the modification required by building laws and fire regulations.[16]

The department now had to obtain approval from Treasury Board to spend the funds specified in the architect's estimate. But Treasury Board balked at this urgent request for funds, feeling it was being "stampeded in arriving at a decision."[17] It demanded to know more of the goals and nature of the Gold Rush Festival. And, as the deputy minister, Robertson, summed up its response, "[m]ost important, the Board wanted to know the justification for the work being undertaken by the Parks Branch. It appeared (to either the Board or its staff) that the auditorium restoration would be a historic restoration put up under the guise of a tourist spectacle."[18] But Robertson silenced further meddling by reminding Treasury Board that the prime minister himself had called for the project,[19] and so the money was approved.

In eliminating all obstacles in the way of the minister's request that the building be ready on time, the branch bent to the task with an unusual degree of goodwill and harmony. The historic sites division did not raise the issue that a national historic site was being desecrated but concentrated its efforts to ensure that the reproduction would be as accurate as possible. The engineers welcomed the co-operation of the historical side of the branch and included it in the planning and development stages of the project. While the architects felt confident about reproducing exterior details, they were less sure about the interior of the auditorium, which had undergone many changes over the years, and left the historic sites division to concentrate on this problem. The division advertised in theatrical trade journals and elsewhere for photographs and descriptions of the theatre as it existed at the beginning of the century. A researcher was hired to visit libraries and archives in Vancouver, Seattle, and San Francisco, and a staff member searched for suitable antiques with which to furnish the building. The Palace Grand Theatre, as it was now called, was built on time and was an immediate success as a tourist attraction and focal point for a heritage development.

But in plunging into the reconstruction of the Palace Grand, the branch became mired in a large and unfocused project to develop Dawson as a tourist attraction. In this regard Treasury Board was right to be nervous about the larger implications. The minister had already pledged that the SS *Keno* would be moved from Whitehorse to Dawson and possibly refurnished as a museum. Discussions were taking place about the department restoring other buildings in the town such as the former cabin of Robert Service. Of immediate concern was that there was no infrastructure of hotels and restaurants to support the tourism brought to Dawson by the festival and the Palace Grand. Côté conveyed this concern to the branch in his report of a senior management meeting.

> An extensive discussion revolved around the fact that a $300,000 restoration would be meaningless and more than wasteful in a community of 300 unless strong leadership is given to ensure that, by 1962, facilities are available to feed and lodge the actors and participants in the Gold Rush festival. It was thought, initially, that the Historic Sites Division had a responsibility to spark this development. As a result of this discussion, the group concluded that the provision of facilities had to be developed *locally* in the cadre of historic buildings still remaining in Dawson.[20]

Avoiding responsibility for the entire tourist development at Dawson became a major concern of the branch in the 1960s.

The department's involvement in Dawson began as a co-operative endeavour and, although carried out on a much larger scale, in this respect resembled co-operative projects undertaken in the 1950s. Unfortunately, for both the people of Dawson and the parks branch, the Dawson Festival was a failure. The place was just too remote, the attractions too limited, and the tourist services too primitive for it to succeed like a Stratford or a Niagara-on-the-Lake. For a few years in the mid-1960s the town tried to promote Dawson as a tourist attraction without government assistance through the Klondike Visitors' Association which rented the stern-wheeler *Keno* and the Palace Grand from the parks branch. But by 1966 it was clear that this venture too was a failure. The town lacked the revenue to develop its attractions, and the parks branch was concerned that activities not in keeping with the gold rush theme were being carried out in the auditorium and the vessel.[21] As a result, the historic sites division investigated the possibilities of taking charge of the operation, and in 1967 the board met in Dawson to review recommendations for further departmental involvement in the area. The board recommended the acquisition of a number of Dawson buildings, urged the protection of others, and advised that a commemorative program be implemented to illustrate the history of the Yukon gold rush. In short order the historic sites division initiated a large research and development project that was then carried out under the area superintendent and the park's regional office in Winnipeg. Subsequently, as the megaproject that became Dawson National Historic Park evolved, neither the townspeople nor the board had much involvement.

As with the Palace Grand, the decision to reconstruct the Fortress of Louisbourg stemmed from larger concerns of the Diefenbaker government for regional economic development. It, too, was a hastily conceived plan which was given an extraordinary timetable. But here millions of dollars, not thousands, were to be spent, and here there was fierce internal rivalry in the branch.

Serious discussions for the partial reconstruction of Louisbourg really commenced in 1958. Cape Breton groups mobilized to celebrate the bicentenary of the siege of Louisbourg and proposed that the event be marked by a partial reconstruction of the fortress. The Associated Boards of Trade of Cape Breton presented a "Twenty-two Point Development Program" to the federal government which, among other things, called for a ten year plan to reconstruct a section of the King's Bastion, including the chapel, a representative section of the hospital, and a typical dwelling, and to enlarge the museum.[22]

The branch development plan that went to the minister, Hamilton, at the beginning of 1959 cautiously approved the idea of a partial reconstruction: "We might reconstruct the Dauphin Half Bastion, the smallest of the bastions, but perhaps the most significant as the principal entrance to the town. Alternatively, and at less expense, we could reconstruct the entrance gate alone, at the Dauphin Half Bastion, which would by itself convey a vivid idea of the architectural significance of the town."[23] However attractive the idea, it was proposed merely for discussion, and no funds were requested for reconstruction.

The idea of reconstruction was frequently debated from 1958 to 1960. The engineers discussed its feasibility.[24] The archaeologist sent to Louisbourg to prepare a largescale investigation was in favour of a partial reconstruction. Katherine McLennan, the honorary curator and daughter of the man who first proposed the idea half a century before, also preferred reconstruction. But she spoke cautiously of this plan, favouring a small rather than a large project. "As to the future," she wrote Herbert, chief of the historic sites division, "the poor climate, the short season, and being at a 24 mile dead-end make planning difficult, the most grandiose scheme would be to rebuild the Citadel [Bastion du Roi] ... refurnish the Governor's quarters ... refurnish the chapel and use the half which contained the barracks as a museum."[25] So, although by 1960 there was a lot of talk about reconstruction at Louisbourg and although most people agreed that a partial reconstruction would be the best solution, there was not a lot of optimism that largescale development would soon be forthcoming.

This situation changed dramatically during 1960 with the appointment of Mr Justice Rand to a one man Royal Commission on Coal. The Cape Breton coal mines had closed down, and Justice Rand was commissioned to investigate options for economic development for the region. Interested in tourism as a possible aternative to coal mining, Rand examined the possibilities of park or historic site development in the area. He contacted officials of the parks branch, including Herbert, and asked them what they "considered would be an appropriate program of development should an unexpected windfall of funds be made available for use in Cape Breton. He hoped that it would provide a substantial works program which would result in employment for a large number of men for several years, and that we would bear in mind the desirability of establishing a continuing tourist industry in Cape Breton Island based in part on the potential of Louisbourg."[26] For Rand, Louisbourg provided an ideal solution as a labour-intensive project that would stimulate secondary industry.

Parks officials were asked to provide technical answers to a hypothetical question: What would be involved in the reconstruction of Louisbourg in terms of money and manpower? In preparing their answer the division chiefs traveled to Williamsburg and Fort Ticonderoga and consulted individuals in the United States and Canada as well as drawing upon their own experience. The Rand Commission report, made public in August, accordingly recommended the reconstruction of Louisbourg and the development of Cape Breton Highlands National Park, and "that both projects be planned in substantial dimensions to extend over a period of from 15 to 20 years during each of which not less than appproximately an expenditure of $1,500,000 will be contemplated."[27]

Even before the tabling of the Rand Commission report the parks branch began to seriously approach the question of a major development at Louisbourg. At a planning meeting held in May, J.R.B. Coleman, the branch director, informed his staff that it was "likely as a result of Rand Commission recommendations that very large sums in the region of $15,000,000 or upward might be made available for the restoration of Louisbourg."[28] The question was how would the branch spend this money. At the meeting that Coleman addressed there were two sides on how to proceed. "On the basic question as to a little restoration and re-creation and a good interpretive program, or a large and extensive re-creation and restoration and a good interpretive program, there were distinct differences of opinion."[29] These opposing views coalesced around two rivals in the National Parks Branch, the historic sites division, headed by Herbert, and the engineering services division, headed by G.L. Scott.

Herbert, with J. Russell Harper, the archaeologist who had been working at Louisbourg in 1960, favoured a partial restoration illuminating several periods of the site's history, including the destruction of the fortress by the British in the 1760s. The engineering division, on the other hand, preferred a total reconstruction of the fortress to resemble its appearance at a single period before its fall. A number of issues were attached to this debate: Herbert did not favour obliterating the ruins and a for a while was opposed to even a partial reconstruction, while the engineers were attracted to a great engineering challenge and were aware of the tourist potential of a Williamsburg of the north.

But the debate focused on one issue: what historical lesson would be drawn from the site? Herbert "felt that an overall re-creation of the fortress as it was in a certain year would not constitute a sound interpretation of history. This would be static in that it would show only the situation and the scene at a certain time – it would not demonstrate the process of history and the changes wrought by time and the fortunes of war."[30] For Herbert, the significant historical event connected with Louisbourg was its fall, and any development should not obscure this fact. The engineers disparaged this approach. One of Scott's staff who had consulted with Rand reported that: "I have always felt that the great significance of Louisbourg as a national monument, was not the conflict between the French and the English but that rather it was a glorious example of the courage, resourcefulness and faith that men had in the new world. To me it is more a study in human history and would be more of a monument to such than a monument to a conflict between two nations."[31] This became the engineers' creed.

The battle over the development of Louisbourg was fought on the ground of department policy. Having received the green light to proceed with the reconstruction, the department had to prepare a development plan with a budget for cabinet approval. Normally a document of this kind would propose two or three options, but the choice would usually be clear. Whoever controlled this submission could affect future policy.

At first it seemed as if Herbert had the upper hand. His proposal was more firmly attached to the twelve million dollar limit imposed by the government. In contrast, the engineers, who admitted that the total reconstruction they favoured would cost in the neighbourhood of forty million dollars, not including interior furnishing, were reduced to proposing a truncated version. As Herbert could promise a total concept for twelve million dollars, he had the advantage. Further, it seems that he was able to convince senior management of the soundness of his interpretive scheme. The deputy minister, R.G. Robertson, told his assistant, Côté: "It seems to me that we ought not to think in terms of a homogeneous level of restoration or partial restoration over the whole area. This would be monotonous, not very effective, and it would also run counter to the very sound point you made about maintaining in part the utter destruction that was imposed on the Fortress."[32] So it was that Herbert was able to draft the cabinet document on the Fortress of Louisbourg. Given free rein to present his own views on how to treat the site, he even managed to subtly rebuke the engineers: "So exciting is the technique (of historic reconstruction) that one can lose sight of history itself. Thus, the desolation of Louisbourg today must not be completely eliminated, otherwise the lesson of history will be lost."[33]

For their part, the engineers were not without means to secure their proposal. As the operational arm of the branch, they had considerable influence over the director, Coleman. Consequently, the director was won over to their camp and lobbied Robertson on their behalf: "A plan for the complete restoration of Louisbourg would bring to the Branch, to the Department and to the people of Canada world-wide attention with resultant publicity and acclaim both inside and outside Canada which would surpass anything that the United States through the Rockefeller Foundation has enjoyed by its efforts at Williamsburg." Realizing that political expediency would not support such a massive project, Coleman suggested that a partial restoration along the lines proposed by the historic sites division would be agreeable providing it would "be planned and carried out such that we always have in mind the ultimate restoration."[34] The alienation of the historic sites division from the branch was indicated by the contrasting sensitivity shown by Coleman to his natural parks. For, after arguing for a massive development of Louisbourg, he counselled against doing very much at Cape Breton Highlands National Park. "We could do a lot of damage to Cape Breton Highlands with a hasty heavy program of work ... I think an experienced engineer and a planning officer could spend at least one and possibly two years full time in Cape Breton making the necessary investigations before we spend a dollar."[35] No such compunction curtailed his enthusiasm for the Louisbourg plan.

The department's proposal for Louisbourg was submitted to cabinet in March 1961. The project was given general approval and $1.1 million allocated for a crash program of preparation to be carried out that year. In March 1962 cabinet selected the option for a partial reconstruction at a cost of $12 million to be spent over twelve years. Cabinet stipulated that a fairly spectacular showing must be made by centennial year 1967 by which time $8.75 million would have been spent.[36] Construction commenced in 1962, very soon after cabinet approval.

Ironically, although Herbert had greatly influenced the cabinet submission, he completely lost control of the subsequent development. Cabinet, it seems, was not concerned about nuances of historical interpretation; it wanted a spectacular showing for 1967 for under twelve million dollars. The engineers, meanwhile, took control of the means of production, completely excluding Herbert and the historic sites division in the process. The collaboration of the branch director, J.R.B. Coleman, was instrumental in this success as he had ultimate responsibility for the project and tended to favour Scott over Herbert.

Upon receiving the preliminary cabinet approval in 1961, the first task was to prepare a detailed development plan and work schedule. Normally the historic sites division would have played a significant role in this stage as it did at Dawson, but the entire planning excercise was contracted out to Ronald Way, a heritage consultant. Way was the foremost expert on heritage development in the country; he managed the restoration of both Fort Henry and the creation of Upper Canada Village. He had both engineering and research experience and knew the problems involved in undertaking large projects.[37] He seemed to be given a relatively free hand in designing the project, but as he reported to Coleman rather than Herbert, his report reflected the approach of the engineers.

Way and his wife Beryl, who acted as his assistant as she had at Upper Canada Village, studied the problem in the summer and presented their completed report in September 1961. Way began by attacking Herbert's interpretive scheme. He categorized two types of restorations, simple and complex. The former focused on a single historical point, while the latter related to multiple points in time. It would be simpler and more effective, argued Way, to restore the fortress to its appearance at a single point in time. Moreover, Way did not agree that the destruction of Louisbourg, and the conflict which brought about its fall should be the central historical theme presented at the site. Under the heading "Educational Objective," Way "recommended that the message of restored Louisbourg should be the story of the progress of Canada's two major races from armed hostility to their national partnership and unity in the Canada of today." Elsewhere he noted that "[t]he Louisbourg restoration offers an absolutely unique opportunity for the visual presentation of a cross-section of the social life of 18th-century New France." Rebutting Herbert's argument that restoring Louisbourg to its appearance before its fall would be ahistorical, he said that "[a] British monarch's decision to obliterate Louisbourg can not be erased from history by a present day decision to restore Louisbourg for a high national purpose. With the perspective of time, it is not inconceivable that Louisbourg's restoration may bring about the fortress' greatest contribution to Canada."[38] Way was not just in the engineers' camp, he had emerged as a powerful opponent in his own right.

Way was still bound by the government policy advocating a partial restoration, and the practical aspects of his report dealt with this question. Generally, he proposed restoring the land side of the fortress to its appearance in 1758 as this would be the side of approach by the tourist, while "the ruined desolation of those on the seaward side would ... have maximum impact on the visitors' imagination."[39] His "phased development plan for research and restoration," however, focused on the reconstruction. He proposed a six year plan in keeping with cabinet's desire to make a significant showing for the centennial year. Fiscal year 1961–2 would initiate the "crash program," which would build a work compound, train semiskilled stonemasons, and initiate manuscript and archaeological research. Limited restoration and reconstruction would be carried out during the following year, while planning and design would commence to produce working drawings for "the reconstruction of major buildings such as the Château St-Louis, the hospital or the Intendant's Palace."[40] These buildings form a complex now known as the Bastion du Roi. Reconstruction of principal buildings was scheduled to begin in fiscal year 1963–4. During the following year work was to commence reconstructing the fortifications on the land front, including the Dauphin Gate, while proceeding with the reconstruction of the Château St Louis and the hospital. The British siege camp was planned to be created in this year, as well as a start made "on the reconstruction of minor buildings within the town ..."[41] During 1965–6 the Dauphin Gate and hospital were scheduled to be completed, and a beginning was made on fitting out their interior furnishings. Construction of a new museum and visitor reception centre also would begin in this year.

At the time that Way submitted his report, Herbert was not completely out of the planning process. In a detailed memorandum to the director, he criticized Way's interpretive scheme.

> Can we allow the challenge of restoration or even "its effectiveness in the field of education" to blind us to the significance of the site in history? This significance may be expressed simply; "France staked her new world empire on the defences of Louisbourg and lost." Surely this is our first message. We can reveal much of social, cultural and military significance while relating the above story, but it will be incidental to the main theme. If visitors leave the "restoration of Louisbourg" without having stood in awe of history, we will have failed. Something of that atmosphere now hangs over the neglected ruins, and we must not destroy it, but enhance it.[42]

Having criticized Way's interpretive scheme, Herbert went on to argue against his proposal for an almost complete reconstruction of the landward fortifications including the citadel and the Château St Louis. But Herbert had very little leverage on the project, and his criticism only served to isolate him even more. Subsequently, he became almost completely powerless to influence interpretation and development at Louisbourg.

At a meeting held in Côté's office attended by the Ways, Scott, and Herbert it was agreed to accept Way's "restoration program" with modifications. The meeting agreed with Way's recommendation that the fortified walls on the landward side should be reconstructed, but not to the extent that he proposed. The reconstruction would not include the Queen's Bastion, and "the balance of the walls would be left 'as is' to denote the desolation and ruin which the British carried out after the fall of Louisbourg."[43] The King's Bastion was to be completely reconstructed, including the chapel and the governor's office, but not the hospital. These activities, then, defined the project that was to preoccupy the branch for the next five years.

The branch proceeded with the Louisbourg project as if the historic sites agency had never existed. Although the work was formally under the supervision of Coleman, the branch director, this authority was delegated to the chief of engineering services division, Scott.[44] Key positions were filled by men personally selected by Scott, including that of the project manager, A.D. Perry, formerly a parks engineer resident at Halifax.[45] Ronald Way was retained as general consultant and, although resident in Toronto, continued to play a key role in the early phases of the reconstruction. He reported to Scott, not Herbert. Even the research staff hired to conduct archival and archaeological investigations reported to Scott through the director of research, Fred Thorpe, hired in 1962. Thorpe's Ottawa staff was kept separate from Herbert's fledgling research unit and worked out of the director's office.

Not just the historic sites division was bypassed; so too was the Historic Sites and Monuments Board. It was not consulted at any point in the planning process, either formally or informally and at its 1962 meeting it passed a rather pathetic motion expressing its concern: "Louisbourg being one of the most important historic sites in Canada, the Historic Sites and Monuments Board of Canada expresses concern over the fact that its partial restoration has been undertaken without consulting the board, and urges that adequate precautions through proper historical and archaeological research be taken to ensure the integrity of any restoration."[46] But like Herbert, the board found itself unable to implement or even express its views.

As development got underway in 1962, a major conflict emerged over the relationship between research and development. Engineers were in charge and they had little understanding of the nature of historical and archaeological research. The importance of initiating research well in advance of construction was not sufficiently appreciated by the engineers, and Thorpe's staff was chronically undermanned. When the researchers announced that their work was not complete enough to allow reconstruction to begin as scheduled, the engineers blamed those impractical people for holding up the project and announced that work would proceed whether they were ready or not.

Disagreement centred on the reconstruction of the King's Bastion which was scheduled to go ahead in the fall of 1962. That summer a team of two archaeologists and six student assistants had directed a work force of forty-five men carefully excavating the area to reveal the foundations of the original complex. At the same time contract historians searched archives in London, Paris, and Ottawa for descriptions of the original fortifications. The researchers did not feel that their investigations were complete enough by the fall to allow the construction of an accurate replica, yet Perry, the project director, who had already fallen behind his work schedule because of bad weather and labour difficulties, was anxious to press on to the next stage. With Herbert out of the picture, it was left to Ronald Way to mediate between the two sides.

Ronald Way sent a memo to Scott in October drawing his attention to the atmosphere of "frantic haste." He went on to say that "[t]his very hurry disturbs me even more than the possibility of restoration not being commenced this fall. I am convinced that while one can use a 'crash program' to build modern workshops, roads and houses, we are asking for trouble when restoration is rushed to the point where no one has time to evaluate research findings carefully, the final archaeological report has not even been prepared, but proposals involving basic restoration policy must be cleared overnight and approved by wire."[47] Another problem that Way pointed out to Scott was that there were not enough stonemasons ready to do a competent job. This last point convinced Scott to delay work on the King's Bastion and he advised Perry accordingly.[48] But Way had identified a chronic weakness of the project, one that got worse instead of better. In 1963 Way attempted to explain the nature of the problem to Scott.

> The essence of current research difficulties is that the ideal situation whereby archaeological and manuscript research is a year or more ahead of design work does not exist with our project. Every one is, in effect, treading on someone else's tail. In my opinion this has come about for three reasons: firstly, the regrettable delay in selecting and hiring the research team, secondly the fact that the research section has never been up to authorized establishment and thirdly the sheer mass of manuscript evidence which must be sifted and analyzed.[49]

But this merely explained the situation, it did not correct it. Way preached compromise to both sides, yet they grew further apart. The engineers became more impatient with the researchers, and Thorpe's comments were increasingly moving in this direction: "it is nonsense to suppose that you can decide what you are going to construct, by what date, and how you are goint to interpret it, and then expect historians and archaeologists to provide all the necessary information when you want it."[50]

With the engineers in charge, the position of the researchers was weak to begin with. It was further hindered by the help they received from their friends, Ronald Way and Jack Herbert. Data for historical reconstruction can be taken from three sources: archaeology, documentary research, and comparative examples. Thorpe's staff had become mired in the first two areas. While sympathetic to their problems, Way embarked on a search for similar examples to Louisbourg in Europe. The Ways toured Europe during the fall of 1962 and brought back an optimistic report about the possibilities of the comparative line of investigation. Assistant deputy minister Côté remarked: "What struck the Ways most was that France contains at Brouage and Mont Dauphin two fortresses of the same vintage as Louisbourg, done by Verrier and intact. Because Louisbourg was razed and we have not got plans for all buildings (and indeed none for the second and third floors of the Château St. Louis) it will be essential to go to 'the typical'."[51] What Côté was saying was that if they could not get the information any other way, then they could look to the typical. The engineers, however, interpreted this to mean that they did not have to wait for historical and archaeological research to be completed but could proceed according to Way's archetypal examples. Thus Scott wrote over the director's signature: "Rightly or wrongly, Mr. Scott is inclined to place more confidence in Mr. Way's basic knowledge of fortress design and the existing fortresses in France than in documentary research."[52]

Similarly, help that research staff received from the historic sites division was couterproductive. In 1962 the newly appointed staff archaeologist at Ottawa wrote a strongly worded memo to Herbert in support of the Louisbourg researchers. Speaking generally of development at historic parks, he remarked:

> Presumably national historic sites are so designated because of their importance in this country's history. It would seem to follow, therefore, that the archaeological remains to be found at such sites are of national historic importance and worthy of protection and preservation ... Pot hunters and amateur archaeologists are extremely destructive of archaeological sites, but the damage done by these diggers is negligible compared to the destructive potential of the bulldozer.[53]

It did not take much imagination to see this as an attack on the engineers' work at Louisbourg. Yet by accepting the historic sites division as an ally, the research staff identified itself with an old foe of the engineers. The engineers already distrusted Herbert, and this action on the part of the researchers would have been regarded as almost treasonous.

At a meeting held at Louisbourg in October 1963 with the research staff and engineers this issue came to a head. Scott and Herbert both flew in from Ottawa in an effort at team building, but Scott was very much in charge, and the hegemony of the engineers was effectively established. Herbert gently tried to establish the integrity of research by asking "what takes precedence and sets the pace: research or construction?"[54] But Scott was unequivocal about the primacy of construction, saying: "'the Deputy Minister had advised him that restoration must go on' despite the inability of research to keep up."[55] Next the senior archaeologist of the project pointed out that "reconstruction, in order to be authentic, should be based upon the findings of the Research Section and constructed in the manner outlined in the reports."[56] Still, the chief engineer was unmoved stating, "that this was not so – that the task of the Research Section was to produce reports to the best of their ability and their responsibility ended with the submission of the reports. From there on it is a matter of policy as to what is done and this is not decided at the Section level."[57] This attitude hardly cleared the air, and bad feelings, like the fog, became endemic at Louisbourg.

It would be wrong to assume that historical and archaeological research was disregarded at Louisbourg. It was not. During the 1960s the research staff grew at Louisbourg, augmented by summer help, and a considerable body of data was accumulated to guide the building of an authentic reproduction. In 1964, for example, a researcher was hired to aid in the search for authentic examples of period furniture for the interior refurbishing of the reconstructed buildings. Archaeological and documentary research became more closely associated with the site after the research section was formally located at Louisbourg in 1963.[58]

In a sense 1963 represented the absolute nadir in the fortunes of the researchers. Perhaps the engineers had gone too far because after this point their power began to wane, and that of the historic sites division grew. Authority for Louisbourg was transferred to historic sites in 1965. Before that Herbert himself had recruited John Fortier from the Royal Ontario Museum, the man who became the national historic park's first professional superintendent. In 1967 Ernest Côté, who had become deputy minister, described Louisbourg as the making of the program: "[t]he project has in many ways been a thorn in the administration's side but in the long term it has been the essential proving ground which has led the Canadian Historic Sites system from its modest position as a weak relative to the Parks side of the operation to its role of partnership with the latter side of the Branch."[59] The division was conscious of its new status as the equal of the national parks agency, and its officials referred to its regeneration as coming "out of the trauma of Louisbourg."[60] The Fortress of Louisbourg became the centrepiece of a distinct network of national historic parks, defining the whole as Banff defined national parks.

But the sovereignty of the historic sites division over this large piece of real estate was more apparent than real. For one thing, a process of regionalization had placed the historic park under the jurisdiction of a regional office in Halifax, and this subdivision integrated the national park and national historic park components of the branch much like the earlier headquarters organization. Operationally orientated, the regional office was also dominated by engineers. It would take a few more years before this imbalance was corrected.

The power of the historic sites division to determine the course of development at Louisbourg was further restricted by the autonomy of Fortier, the park superintendent. The superintendent's great strength appeared to be his ability to get along with people in the poisoned atmosphere of Louisbourg, and his enthusiasm and vision rejuvenated many who were tired of years of internecine rivalry. His interest in protecting high standards of authenticity while his promotion of a successful tourist attraction drew both sides to him. But it soon appeared that the superintendent had his own agenda for reconstruction, one that was different from that agreed upon in Ottawa, and his smooth ways and silver tongue were suspected of being used to circumvent head office policy.

The superintendent favoured interpreting the social history, rather than the military history of the site, and proposed reconstructing portions of the original town. At the staff meeting in 1963, for example, amid all the negativity and ill-feeling, he stood up and "said that to create a total atmosphere of living history it was his opinion that we should consider the possibility of totally constructing blocks of housing so as to have one part of Louisbourg as completely reconstructed as possible."[61] At the time this suggestion was dismissed out of hand because of the costs, but there was a great deal of obvious sympathy both from Way and the engineers. Gradually the idea grew more real, and by 1966 the possibility of restoring part of the town was accepted in the development plan as a potential but not probable line of work.

> Certain private dwellings will be developed to depict the life and times within an eighteeth century fortress, and the main quay area and wharves of the Fortress will be developed since they played such an important part in the military and trade history of the Port of Louisbourg. It is considered that the private and public buildings referred to above should

include virtually total reconstruction of the three blocks that lie west of the road on which the present museum stands. ... At the present time this latter proposal is approved for pre-planning purposes only.[62]

Whatever the status of this proposal, it greatly influenced the interpretation of the site. The engineers' vision to have Louisbourg represent the social life of New France to the exclusion of its defeat was completely triumphant. What concerned John Nicol, the new director of national parks, however, was that the superintendent appeared to be acting on this barely official proposal without proper authorization. Thus in September 1966 he informed the assistant deputy minister that "it would appear that the Park Superintendent is trying to manœvre the Department into a position of opening up many areas of work which it would have to complete no matter what ceilings were proposed."[63] By 1967 the original plan had been abandoned, and reconstruction entered a new phase apparently guided more by the agenda of Fortier than by headquarters.

By introducing the big project to regional development, the federal government placed undue emphasis on historic parks. As a result, the smaller co-operative endeavours tended to be forgotten as the administration strove to deal with large projects such as Louisbourg and Dawson. In the process the new status of the advisory board was lost, and it reverted to its old concern of "getting the plaques up." With the board relegated to the perimeter, the program lost an important tool for historical interpretation. Without the regional perspective of the board, the historic park developments were prone to lack a point of view as they strove to achieve a chimerical national perspective. As a journalist wrote recently about the flagship of national historic parks: "Today Louisbourg stands as a symbol of our national puzzlement over what history is truly Canadian and how to possess it."[64] Still, if national historic parks cannot present a unified image of the Canadian past, there are few historians and fewer cultural institutions that can.

BEYOND THE 1960s

The era of the big project continued through the 1970s, and the historic park system was enlarged from coast to coast. In Newfoundland the site of an early Viking settlement was excavated between 1973 and 1977. The property was developed as L'Anse aux Meadows National Historic Park, and a replica built of one of the Viking houses.[65] In Prince Edward Island the program undertook the restoration of the room in Province House where the Fathers of Confederation met at the Charlottetown Conference and developed the story-book setting of *Anne of Green Gables.* In Nova Scotia the Halifax Citadel and the Fortress of Louisbourg continued to receive considerable attention during the decade as millions of dollars were poured into restoration and reconstruction. A substantial restoration project was undertaken at Les Forges St-Maurice National Historic Park at Trois Rivières, Québec in 1973. In Ontario the Rideau and Trent canals were incorporated into the national park system as recreational areas, and considerable work was done to identify, preserve, and interpret their heritage features. At Lower Fort Garry National Park in Manitoba restoration and reconstruction begun in the 1960s was continued. In Saskatchewan development work was carried on at Fort Walsh which had been acquired from the RCMP in 1968. A stockade was constructed to replicate the one surrounding the original fort, and a visitor reception centre with costumed guides was established. Alberta still remained weakly represented, although the department initiated development at Rocky Mountain House. Meanwhile, British Columbia received further bounty: Fort Rodd Hill was established as a national historic park and Fort St James National Historic Site also was acquired and restored. But the largest development outside of Louisbourg was at Dawson where the department acquired and restored a number of buildings to complement the reconstructed Palace Grand Hotel. By the 1970s Parks Canada was the leading employer in the town and the mainstay of the area's tourist industry.

The organization became both large and decentralized as it served these developments. Operations at the various historic parks were managed by the superintendents who had, since the late 1950s, been recruited as regular civil servants with middle management classification. They looked after ongoing maintenance and administration of the parks. Large developments such as Louisbourg and the Halifax Citadel had their own core of professional staff including archaeologists and historians. Otherwise technical assistance for new development was handled either in the regional office of Parks Canada, or at branch headquarters in Ottawa.

The administration of national parks had become decentralized in accordance with the Glassco Royal Commission's recommendations of the late 1960s. By the mid-1970s the administration of national parks, including historic parks and sites, was delegated to five regions with offices in Halifax, Québec, Cornwall, Winnipeg, and Calgary. Each of these centres established a research function with their own staff of archaeologists, historians, and interpreters. This enabled them to look after the short- and medium-term development of heritage sites in their respective regions but caused some difficulties as the regional offices integrate the national and historic park activities into a single program much as they were at headquarters in the 1950s. As before, there is the tendency for the outlook of the regional offices to be dominated by the concerns of the national parks. More recently this tendency has been partly corrected by the advent of functional management where directors of specialist branches in Ottawa review regional initiatives in their areas of expertise.

At headquarters, the organization of the historic sites program mushroomed in the 1970s into a self-sufficient agency having separate divisions looking after research, conservation, interpretation, and planning. Its operational capacity was further enhanced by a restoration services unit closely attached to its program. The headquarters organization assumed responsibility for medium- and long-range development at historic parks, as well as providing support for the Historic Sites and Monuments Board. Through the 1970s senior management tended to be exclusively preoccupied with park development and paid little attention to cost-sharing agreements or architectural conservation. The various components of the Historic Parks and Sites Branch reflected this orientation and focused their attention on improving old and newly established historic parks.

The board meanwhile found itself alienated from these big projects. In the first instance decisions to proceed with megaprojects such as Louisbourg or Dawson or even with more modest developments such as Bellevue, the former home of Sir John A. MacDonald, were too often taken without the board's advice. Secondly, the decision on how to develop these properties was left to the growing number of experts who ran this part of the program. Whether academic or amateur, the board members were no match for the technical apparatus of the department in these affairs. Even planning had been taken over by professionals within the service. Excluded in these ways, there was little left for the board to do except meet twice a year to consider nominations for national historic sites. It was in these circumstances that the board drifted back to its old preoccupation with commemoration.

A policy of government restraint beginning in the late 1970s and continuing into the 1980s has put an end to the era of the big project. Work at large historic parks such as Louisbourg, Halifax

Citadel, Lower Fort Garry, and Dawson has been largely curtailed, and the agency is not contemplating any major new projects in the near future. This has been the case for several years now, and the situation is beginning to transform the mindset of the organization. It would seem that both senior and middle management are no longer preoccupied with historic parks and are returning to concerns more compatible with the new fiscal atmosphere. In its 1982 publication, *Parks Canada Policy*, for example, the department stated its intention to develop new initiatives in the area of architectural preservation. Its objective was "to act as the coordinating federal agency in fostering the protection of Canada's architectural and cultural heritage through: the conservation of heritage buildings under federal jurisdiction; the elimination of disincentives to heritage building conservation; the development of cooperative programs with the provinces and territories to encourage public and private initiatives in this field."[66] This new direction has been slow to manifest itself, and of the proposals suggested in this part of the 1982 policy statement only the Federal Heritage Building Review Office has been established. Nonetheless, the path of future activity is clear.

Meanwhile, other players have emerged to influence the politics of historic sites. Most of the provinces now have fully-fledged heritage programs of their own, and have undertaken comprehensive policies for archaeological, architectural, and historical preservation. The recent trend has been for these provincial agencies to try and co-ordinate the efforts of local preservation groups along the lines of the federal program in the 1950s. Heritage Canada, founded in 1973, has also become involved in organizing grass roots preservation projects. Meanwhile, the trend has moved away from commemoration. Provinces such as British Columbia, for example, no longer plaque historic sites and have abolished the advisory committee.

It is still too early to tell what structural effects this reorientation will have on the federal historic sites administration but they do lead to a number of questions. What effects will the realization that historic parks are no longer central to development have on the deployment of human resources? And what will the end of the era of the big project mean to the Historic Sites and Monuments Board? Will it emerge as an important conduit between public and government initiatives in the area of preservation, or will it slide into obscurity? Only time can tell.

Appendices

Appendix 1

HISTORIC SITES AND MONUMENTS BOARD OF CANADA, MEMBERSHIP BY AREA OF RESPONSIBILITY, 1919–87

I Chairmen

1919–39 Brig. Gen. E.A. Cruikshank, FRSC (Fellow of the Royal Society of Canada), Head of Department of Militia and Defence Historical Section, Ottawa. See Ontario.

1939–42 Vacant.

1943–4 F.W. Howay, FRSC, Judge, Historian. See British Columbia.

1945–50 J.C. Webster, FRSC, Retired Physician, Historian. See New Brunswick.

1950–8 Fred Landon, FRSC, Librarian and Professor of History, University of Western Ontario. See Ontario.

1958–60 Mgr A. D'Eschambault, FRSC, Priest, Historian. See Manitoba.

1960–7 Bruce Fergusson, FRSC, Provincial Archivist of Nova Scotia. See Nova Scotia.

1967–70 Alan Turner, Provincial Archivist of Saskatchewan. See Saskatchewan.

1970–8 Marc La Terreur, Professor of History, Université Laval, See Québec.

1978–81 Leslie Harris, Professor of History and Dean, Faculty of Arts and Science, Memorial University of Newfoundland. See Newfoundland.

1981–5 J.M.S. Careless, FRSC, Professor of History, University of Toronto. See Ontario.

1986– T.H.B. Symons, FRSC, Vanier Professor, Trent University. See Ontario.

II Members – Provincial

Alberta

1944–55 M.H. Long, FRSC, Professor of History, University of Alberta, Edmonton.
1956 M.E. Lazerte, Professor of Education, University of Alberta, Edmonton.
1957–9 Joel K. Smith, Businessman, Edmonton.
1959–67 R.Y. Secord, Rancher, Edmonton.
1968–74 Lewis H. Thomas, Professor of History, University of Alberta, Edmonton.
1975–8 Hugh Dempsey, Director of History, Glenbow – Alberta Institute, Calgary.
1979 Trudy Soby, Heritage Consultant, Calgary (see 1985).
1980–5 Jaroslav Petryshyn, Professor of History, Peace River Community College.
1985 Trudy Cowan, Heritage Consultant, Calgary (see 1979).

British Columbia

1923–44 F.W. Howay, FRSC, Judge, Historian, New Westminster. Also responsible for Manitoba until 1937, and Alberta until 1944.
1944–59 W.N. Sage, FRSC, FRS, Professor of History, University of British Columbia, Vancouver.
1960–7 Margaret Ormsby, FRSC, Professor of History, University of British Columbia, Vancouver.
1967–71 James Nesbitt, Journalist, Victoria.
1971–9 Margaret Prang, FRSC, Professor of History, University of British Columbia, Vancouver.
1979– Charles Humphries, Professor of History, University of British Columbia, Vancouver.

Manitoba

1937–59 Mgr A. D'Eschambault, FRSC, Priest, Historian, St Boniface.
1959–66 W. Smith, Professor of History, Brandon College.
1967–9 E. Russenholt, Businessman, Broadcaster, Headingly.
1969–73 G. Anderson, Merchant, Lac du Bonnet.
1974–5 J.E. Rea, Professor of History, University of Manitoba, Winnipeg.
1976–8 Robert Painchaud, Professor of History, University of Winnipeg (Killed, June 1978).
1980–6 Richard Grover, High School Teacher, Winnipeg.
1987– Vacant.

New Brunswick

1919–23 W.O. Raymond, FRSC, Clergyman, Historian, Saint John.

1923–50 J.C. Webster, FRSC, Retired Physician, Historian, Shediac.
1950–61 A.G. Bailey, FRSC, Professor of History, University of New Brunswick, Fredericton.
1961–2 John Palmer, Lawyer, President, New Brunswick Historical Society.
1962–4 Lt-Gen. E.W. Samson, Retired Army Officer, Fredericton.
1964–9 Gerald Keith, Businessman, Member, New Brunswick Museum Board.
1970–6 G.L. MacBeath, Historian, Deputy Head, Historical Resources Administration, Province of New Brunswick, Fredericton.
1976–8 Jules Léger, Professor of History, Université de Moncton (Killed, June 1978).
1979–87 Jean Daigle, Professor of History, Université de Moncton.
1988– Mrs Marion Beyea, Provincial Archivist of New Brunswick, Fredericton.

Newfoundland
1950–5 C.E.A. Jeffery, Newspaper Editor, St John's.
1956–60 Oliver Vardy, Director, Newfoundland Tourist Development Office, St John's.
1961–6 E.B. Foran, City Clerk, Historian, St John's.
1967–80 Leslie Harris, Professor of History, Dean of the Faculty of Arts and Science, Memorial University of Newfoundland, St John's.
1981–4 Noel Murphy, Physician, Broadcaster, Mayor of Corner Brook.
1985– Shane O'Dea, Professor of English, Memorial University of Newfoundland, St John's.

Northwest Territories
1974–5 Alex Stevenson, Observer, Ottawa.
1976–83 Fr G. Mary Rousselière, Missionary/Archaeologist, Pond Inlet.
1986 Sarah Jerome, Heritage Activist, Fort McPherson.
1987 Vacant.
1988– Mr John U. Bayly, Lawyer, Yellowknife.

Nova Scotia
1919–23 W.C. Milner, Journalist, Archivist, Halifax.
1923–5 J.P. Edwards, Businessman, President, Nova Scotia Historical Society, Halifax.
1925–30 W. Crowe, Judge, Sydney.
1931–54 D.C. Harvey, FRSC, Provincial Archivist, Professor of History, Dalhousie University, Halifax.
1954 Thomas Raddall, FRSC, Historian, Historical Novelist, Halifax.
1955–69 C. Bruce Fergusson, FRSC, Provincial Archivist, Professor of History, Dalhousie University, Halifax.

1970–7 Peter Waite, FRSC, Professor of History, Dalhousie University, Halifax.
1978–87 Raymond MacLean, Professor of History, St Francis Xavier University, Antigonish.
1988– Vacant.

Ontario I

1919–39 Brig. Gen. E.A. Cruikshank, FRSC, Head of Department of Militia and Defence Historical Section, Ottawa.
1939–54 Vacant.
1955–9 Harry Walker, Journalist, Historian, Ottawa.
1959–61 A.R.M. Lower, FRSC, Professor of History, Queen's University, Kingston.
1961–73 James J. Talman, FRSC, Chief Librarian, Professor of History, University of Western Ontario, London.
1973–78 B. Napier Simpson, MRAIC (Member, Royal Architectural Institute of Canada), Restoration Architect, Thornhill (Killed, June 1978).
1978–81 Vacant.
1981–7 Edward Storey, Professor of Recreology, University of Ottawa.
1988– Mr John H. White, Businessman, former Chairman of Ontario Heritage Foundation, London, Ontario.

Ontario II

1919–32 James Coyne, FRSC, Lawyer, Historian, formerly President, Ontario Historical Society, St Thomas.
1932–58 Fred Landon, FRSC, Librarian, Professor of History, University of Western Ontario, London.
1958–72 D.G. Creighton, FRSC, Professor of History, University of Toronto.
1972–85 J.M.S. Careless, FRSC, Professor of History, University of Toronto.
1986– T.H.B. Symons, FRSC, Vanier Professor, Trent University, Peterborough.

Prince Edward Island

1950–8 Thane Campbell, Chief Justice of Prince Edward Island, Charlottetown.
1959–66 Earl Taylor, Merchant, President, Prince Edward Island Historical Association, Charlottetown.
1967–77 Fr Francis Bolger, Professor of History, University of Prince Edward Island, Charlottetown.
1978–89 Irene Rogers, Heritage Activist, Charlottetown.

Québec I

1919–23 Benjamin Sulte, FRSC, Civil Servant, Historian, Ottawa.

1924–5 Victor Morin, FRSC, Lawyer, Historian, Montréal.
1925–6 Aegidius Fauteux, FRSC, Librarian, Historian, Montréal.
1927–9 P. Demers, Librarian, Montréal.
1930–3 Maréchal Nantel, Librarian, Historian, Montréal.
1933–55 E.-F. Surveyer, FRSC, Judge, Historian, Montréal.
1955–60 Jules Bazin, Librarian, Montréal.
1961–9 Marcel Trudel, Professor of History, Université Laval, Québec and University of Ottawa.
1969–78 Marc La Terreur, Professor of History, Université Laval, (Killed, June 1978).
1978–9 Vacant.
1980– Noël Bélanger, Professor of History, Université du Québec à Rimouski.

Québec II

1955–60 Édouard Fiset, Town Planner, Québec.
1960–1 C.G. Dunn, Newspaper Editor, Québec.
1961–73 Jean-Jacques Lefebvre, FRSC, Historian, Judicial Archivist, Montréal.
1974–88 Andrée Désilets, FRSC, Professor of History, Université de Sherbrooke.
1988– Vacant

Saskatchewan

1937–50 J.A. Gregory, Politician, President, Prince Albert Historical Society.
1951–4 Campbell Innes, Regional Historian, Curator, Battleford Museum.
1955–60 R. Mayson, Businessman, North Battleford.
1961–6 A.L. Agnew, Businessman, President, Prince Albert Historical Society.
1967–75 Alan Turner, Provincial Archivist, Regina.
1976–88 David E. Smith, FRSC, Professor of Political Science, University of Saskatchewan, Saskatoon.
1988– Vacant.

Yukon

1973–5 Rev. Ken Snider, Observer, Dawson.
1980–5 Jeanne Harbottle, Heritage Activist, Whitehorse.
1985–7 George Shaw, Retired Entrepreneur, Vancouver.
1988– Vacant.

III Members – Institutional

National Archives of Canada (Ex Officio)

1937–49	Gustave Lanctôt, FRSC, Dominion Archivist.
1949–68	W. Kaye Lamb, FRSC, Dominion Archivist.
1969–84	W.I. Smith, Dominion Archivist.
1985–	J.-P. Wallot, National Archivist.

National Museum of Canada (Ex Officio)

1951–5	F. Alcock, FRSC, Geologist/Museum Administrator.
1956–8	Vacant.
1959–61	Clifford Wilson, Geologist/Museum Administrator.
1963–7	Vacant.
1968–71	W.E. Taylor, Archaeologist, Director.
1971–6	George MacDonald, Archaeologist.
1976–8	James V. Wright, Archaeologist.
1978–	George MacDonald, Archaeologist, Director, Canadian Museum of Civilization.

Appendix 2

ORGANIZATION OF THE HISTORIC SITES BUREAU, 1955[1]

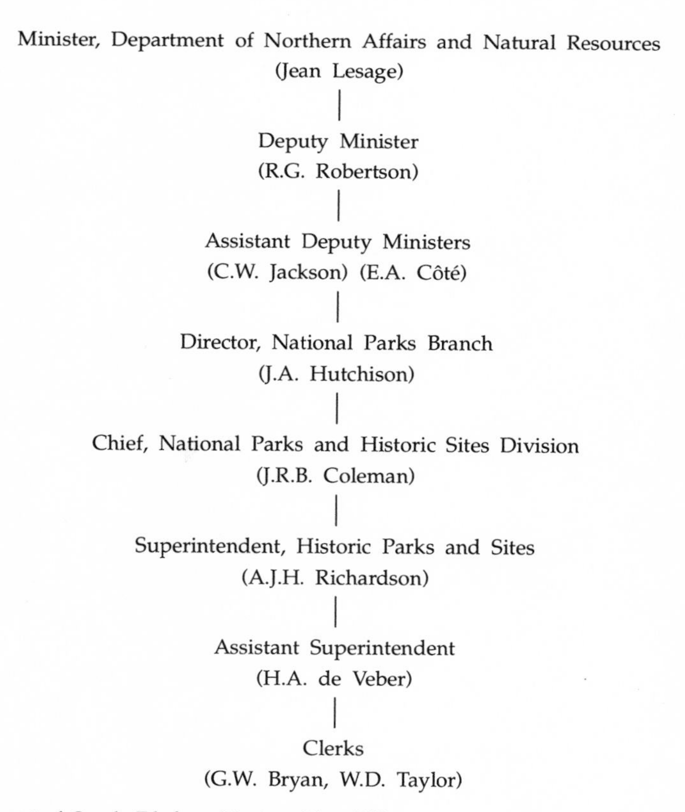

1 *Government of Canada Telephone Directory*, May 1955.

Appendix 3

ORGANIZATION OF THE NATIONAL PARKS BRANCH, 1963

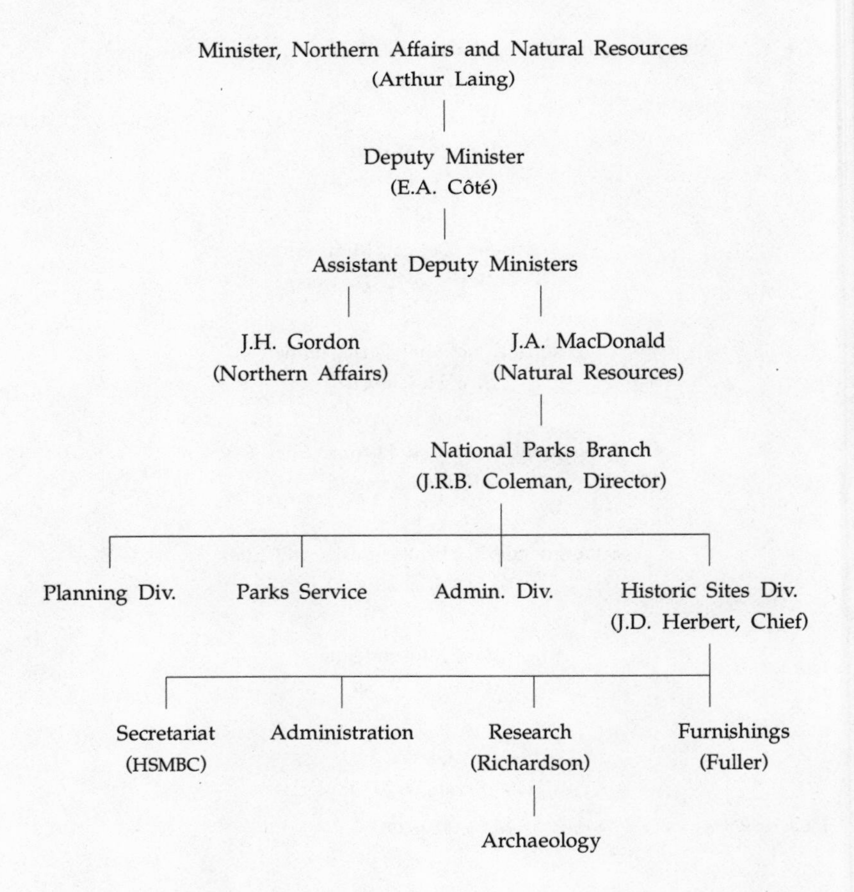

Appendix 4

HERITAGE AGENCIES IN FRANCE, BRITAIN, THE UNITED STATES, AND THE UNITED NATIONS

One of the first effective government programs for the recognition and protection of historic sites was instituted in France. Originating in 1830, it was the result of a centralized government intent on organizing a disparate country, and a growing nationalism which sought to recover national ideals through the celebration of a glorious past. The agency began with the appointment of an inspector general and, in 1837, an Historic Monuments Commission was established. An inventory of historic buildings was instituted, and the first list was published in 1840. Subsequently, a framework of controls and grants was introduced aimed initially at protecting cathedrals and other large public monuments but eventually including a limited number of other buildings. Although granted sweeping powers, the inspector general was hampered by a chronic shortage of funds. A similar situation existed in Italy where listed buildings were endangered by neglect, and in both countries programs were reluctant to extend their authority over areas where they could not offer protection.

This situation was in part remedied in France by the extensive Malraux laws, first enacted in 1966 to better protect aspects of the national heritage. According to American preservation specialist, James Marston Fitch, this legislation has had far reaching implications. "This new legislation created a whole system of grants-in-aid, low interest loans, tax abatements, rent increases ... and technical assistance to enable private owners to rehabilitate listed properties. The consequences of this new policy are to be seen in every town in France ... A similar quid-pro-quo arrangement seems inevitable for any nation which is seriously committed to preserving its cultural heritage, and such a policy is, in fact, being adopted piecemeal in Holland, Great Britain and the United States" (Fitch, *Historic Preservation*, 400).

Preservation in Great Britain typically began in a decentralized and nonofficial way. The first national preservation society, the Society for the Protection of Ancient Buildings, was founded by William Morris in 1877. Another voluntary society, founded in 1894, was the National Trust, which aimed to preserve landscapes as well as buildings. There was no official participation in recognizing and preserving historic sites and buildings until 1882 when the first British Ancient Monuments Act was passed. This act had very little power, principally because it was unwilling to intervene in cases involving private property, and it was not until 1913 when the act was amended that a government agency really came into being. The 1913 act led to the formation of the Ancient Monuments Branch of the Ministry of Works which undertook a national survey of historic sites and preserved a number of historic and archaeologically significant places. Subsequently the Ministry of Housing and Local Government co-ordinated regional inventories of architecturally and historically significant buildings. As elsewhere, however, lack of funding has greatly inhibited the effectiveness of this activity. Since 1970 responsibility for co-ordinating the various national initiatives for conserving and developing heritage resources has fallen on the Directorate of Ancient Monuments and Historic Buildings of the Department of the Environment. According to Robert Hewison, the department spent ninety-six million pounds in 1986–7. "More than half of this," he added, "goes to the Historic Buildings and Monuments Commission, which has adopted the name English Heritage. English Heritage manages some 400 monuments and buildings, makes grants to individuals and organizations to assist with historic buildings, conservation areas, town schemes, ancient monuments and rescue archaeology" (Hewison, *Heritage Industry*, 25).

The national program of the United States bears the closest resemblance to Canada's program. Like Canada, the United States heritage program is largely carried out under the auspices of its national parks service although surprisingly this mandate was not formally established until 1935. Legislation enacted that year established two kinds of heritage property. The first, the National Historic Landmark, closely resembles the Canadian National Historic Site: recognition does not entail public ownership, and the site is usually commemorated by means of a bronze plaque. As in Canada, the selection of these sites involves the participation of an honorary advisory body. The second, National Historic Parks, is also managed by the US National Parks Service. The creation of these parks does not involve the participation of an advisory board but results from the direct instigation of the department. Here, however, Congress has had considerable influence. Following the 1935 legislation, the national parks system acquired a number of battlefields from the War of Independence and the Civil War, and military sites continue to dominate the historic park system.

The US National Parks Service began to expand this activity following the end of the Second World War. The most significant development in the postwar era was the enactment in 1966 of the National Historic Preservation Act. According to an inhouse history of the service, "the 1966 act broadened the Service's concern and responsibilities to encompass properties of state and local as well as national significance" (Mackintosh, *The Historic Sites Survey*, 71). It authorized matching federal grants-in-aid to the states for the survey, acquisition, and preservation of historic properties. The National Register is a co-operative program involving state preservation agencies and the federal parks service. In this program initial surveys, research, and nomination of historic buildings are handled by the state agencies operating under guidelines established by the federal agency. Other legislation passed in 1966 made the granting of federal monies for urban renewal projects conditional on environmental impact studies which had to assess the heritage significance of the area affected.

Concurrent with the National Register, the US parks service has explored other avenues of heritage conservation. One such measure has been the recognition and development of heritage areas. Operating in tandem with state and local bodies, the agency has recognized Lowell, Massachusetts as an historically significant industrial area and taken measures to develop and protect historic and architectural themes within the town. Another initiative coordinates Cultural Resources Management, the protection of historic resources within its extensive natural park system. This program has produced inventories and research leading to the establishment of policies for the protection and proper exploitation of human resources in all national parks.

As in Canada, the United States has a National Trust for Historic Preservation, although the American agency, founded in 1949, has been more active than its Canadian counterpart in chanelling private donations into public preservation projects. What really distinguishes heritage preservation in the United States from that in Canada is the number of large privately endowed projects. Perhaps the greatest of these was the restoration of the colonial town of Williamsburg in Virginia undertaken in the 1920s and 1930s with funding from John D. Rockefeller. The enormity of the task – acquisition of entire city blocks and their restoration to their colonial appearance – was facilitated by the project's initiator, W.A.R. Goodwin, who was provided with virtually limitless resources by his patron. The first phase of this undertaking, completed in 1934, had immense influence on the preservation movement generally, and we have already noted its effect on the advocates for the reconstruction of Louisbourg. Hosmer, in describing it as a great influence on the American preversation movement, has remarked: "It would be no exaggeration to say that by the 1930s the Williamsburg organization had begun to function as an American national trust, as a central clearinghouse for preservation information" (Hosmer, *Preservation Comes of Age*, vol. 1: 65).

Another privately funded heritage project, but one with less of an impact on preservation, was Henry Ford's Greenfield Village, begun in the 1920s. Unlike Williamsburg, Greenfield was not an historic village, but a collection of miscellaneous buildings brought to the site to illustrate aspects of Henry Ford's very personal view of industrial history in America. In this regard it is really an outdoor museum along the lines of the Swedish Skansen or, later, Upper Canada Village.

The United Nations sponsors two heritage agencies through UNESCO: the International Council on Monuments and Sites (ICOMOS) and the World Heritage Convention. ICOMOS was included as a UNESCO participant as early as 1970, although its origins date to the 1960s. It is primarily devoted to publicizing heritage issues such as architectural preservation and has instituted influential guidelines for the treatment of heritage architecture. The most marked manifestation of this latter interest is perhaps the Venice Charter, drafted in 1966. ICOMOS influenced the establishment of the World Heritage Convention.

In 1972, UNESCO launched the Convention Concerning the Protection of the World Cultural and Natural Heritage. The convention aims to recognize landmarks around the world that have international significance because of their natural or cultural associations. Participation is voluntary and to date ninety countries, including Canada, have signed the convention agreeing to protect designated sites within their boundaries and to contribute to the World Heritage Fund which was established to help protect endangered sites.

The sites, selected by committee, include such well-known places as the Egyptian pyramids, the Taj Mahal, and the Grand Canyon. Canada has a number of world heritage sites. Most of these are natural landmarks, but they include L'Anse aux Meadows National Historic Park in Newfoundland and the Québec City historic area (Taylor, "World Heritage Convention").

Notes

ABBREVIATIONS

CHA	Canadian Historical Association
HLAC	Historic Landmarks Association of Canada
HSMBC	Historic Sites and Monuments Board of Canada
NA	National Archives of Canada
ROYAL COMMISSION	Royal Commission on National Development in the Arts, Letters and Sciences
RSC	Royal Society of Canada
UBCL	University of British Columbia Library

PREFACE

1 Hosmer, *Presence of the Past; Preservation Comes of Age.*
2 Cited by Cassirer, *An Essay on Man,* 172. Cassirer argues (p. 185) that history, like life, does not have an objective reality but exists in an objective/subjective synthesis which is revealed through the historical process.

CHAPTER ONE

1 Denison, "The United Empire Loyalists and Their Influence upon the History of this Continent," RSC, *Proceedings and Transactions* (1904): xxvii.
2 Ibid., xxxi.
3 Ibid.
4 Ibid., xxix
5 Berger, *The Sense of Power,* 89.

6 Killan, *Preserving Ontario's Heritage,* 32.
7 Ibid., 4.
8 Hitsman, *The Incredible War of 1812,* 198–9.
9 "One of the most durable of our legends is what I may call the Militia Legend of 1812. By that I mean the idea that during the war the country was defended by 'the Militia' with only a little help from regular troops." Stacey, "The War of 1812 in Canadian History," 154.
10 "Grants for Monuments etc. through the Department of Finance," n.d., NAC, RG 84, vol. 1305, HS 8, pt 2.
11 There is a draft of an inscription in the Denison Papers which suggests that he had a hand in its composition. Denison Papers, n.d., NA, MG 29, E29 vol. 5.
12 *Globe* (Toronto), 29 July 1895.
13 RSC, *Proceedings and Transactions* (1900): viii.
14 Carter-Edwards, "Fort Malden," 290.
15 Ibid., 291.
16 RSC, *Proceedings and Transactions* (1900): viii.
17 Bourinot, "Some Memories of Dundurn and Burlington Heights," in RSC, *Proceedings and Transactions* section II (1900): 2.
18 Killan, *David Boyle,* 99.
19 Boyle, "Archaeological Report," in *Annual Report of the Canadian Institute* (session 1891): 8.
20 Coyne, *Country of the Neutrals,* 1.
21 Killan, *Preserving Ontario's Heritage,* 139. Today, Fort York continues to be preserved on the grounds of the Canadian National Exhibition.
22 Monet, *Last Cannon Shot,* 3.
23 "As early as 1843 Cartier had rejected strident French-Canadian nationalism as well as Durham's argument for the assimilation of French Canada. Canada, he argued, was in a happy 'situation' and 'blessed by providence' to have two great civilizations within its bosom. Both ethnic groups could benefit from the philosophy, history, and literature of the other. English-speaking Montrealers were assured that French-Canadians had British hearts and were committed to the happiness and prosperity of Canada." Young, *George-Étienne Cartier,* 61–2.
24 Roy, *Les monuments commémoratifs de la province de Québec,* volume 1: 13.
25 Two bronze shields, towards which Ottawa had contributed $1,250, were erected in 1904. RCS, *Proceedings and Transactions* (1906): ciii.
26 Guitard, *Militia of the Battle of the Chateauguay,* 83.
27 "The main body of Americans had hardly been involved and their casualties were about 50 officers and men; de Salaberry lost 5

killed, 16 wounded and reported 4 missing." Hitsman, *The Incredible War of 1812*, 166.

28 Guitard, *Militia of the Battle of the Chateauguay*, 88.

29 Sulte, *Histoire des Canadiens–français*, vol. 8: 95.

30 W. Patterson to Mackenzie Bowell, 24 Aug. 1892, NA, RG 84, vol. 1242, HS 7–10 pt 1.

31 In 1921, Sulte recalled: "In 1876 [*sic*] the Privy Council refused to approve a more detailed inscription and I wrote the one now inscribed on the column." NA, RG 84, vol. 952, HS 7–10, pt 1.

32 Thibodeau, "La conservation du Fort Chambly," 10.

33 Mott, "The 'Old Fort' at Chambly," *The Canadian Antiquarian and Numismatic Journal* III, 3 (January 1875), cited in Thibodeau, "La conservation du Fort Chambly," 10.

34 Thibodeau, "La conservation du Fort Chambly" 11. Cf. Brunet, "Trois dominantes de la pensée canadienne-française l'agriculturisme, l'anti-étatisme et le messianisme," in M. Brunet, *La présence anglaise et les Canadiens*, 119; and Monière, *Le développement des idéologies au Québec*, 224.

35 Thibodeau, "La conservation du Fort Chambly," 15.

36 Ibid., 19.

37 Ibid.

38 Memorandum, 20 May 1887, NA, RG 84, vol. 1061, file FC 2, pt 1.

39 Thibodeau, "La conservation du Fort Chambly," 12.

40 Officer commanding Military District no. 6 to adjutant general, Department of Militia and Defence, 25 Jan. 1904, NA, RG 84, vol. 1084, file FLE 2, pt 2.

41 Memorandum from A.A. Pinard to J.B. Harkin, 18 Oct. 1921, ibid.

42 Undated memorandum on the early development of Fort Anne National Park, NA, RG 84, Vol. 1042, file FA 2, pt 4.

43 F.B. Wade to Sir Frederick Borden, 18 Mar. 1904, NA, RG 84, vol. 1041, file FA 2, pt 4; pt 3.

44 Director general of engineers to military secretary, 21 August 1913, ibid., pt 4.

45 Military secretary to A.L. Davidson, MP, 6 July 1914, ibid.

46 *Spectator* (Annapolis Royal), 17 Feb. 1916.

47 Memorandum from J.P. Dunne, superintendent of Ordnance and Admiralty Lands Branch, to F.H.H. Williamson, Canadian National Parks Branch, 26 Nov. 1919, NA, RG 84, vol. 1048, file FB 2, pt 4.

48 Schmeisser, "The Creation and Development of Fort Beauséjour National Historic Park."

49 Parkman, *Montcalm and Wolfe*, vol. II: 56.

50 Williamson, "Report on Investigation of Historic Sites in Maritime Provinces," Dec. 1919, NA, RG 84, vol. 1189, file HS 6 (Maritime Gen-

eral), pt 1. A useful account of early attempts to develop Louisbourg as an historic site is Johnston, "Preserving History."

51 Williamson, "Report on Investigation of Histori Sites," Dec. 1919, NA, RG 84, vol 1189, file HS 6 (Maritime General), pt 1.

52 For Pascal Poirier, who represented the Acadian viewpoint in the Senate, Louisbourg's greatness derived as much from French as British heroism, and in 1895 he rose in Parliament to demand: "Let us raise monuments to the dead without discrimination." Canada, Senate, *Debates*, 1895, 138.

53 Bourinot, "Cape Breton and its Memorials of the French Regime," in RSC, *Proceedings and Transactions*, (1891): 173.

54 "Louisbourg as a National Charge, Splendid Lecture Before Nova Scotia Historical Society by J.S. McLennan," *Halifax Herald*, 11 Nov. 1908, 1.

55 RSC, *Proceedings and Transactions*, (1891): xiii.

56 Ibid. (1901): xxi.

57 Ibid. (1902): xxiv.

58 Ibid. (1903): xxxvii.

59 HLAC, *Appeal for Membership*, 3.

60 HLAC, *Annual Report*, (1915): 9.

61 Sylvia Van Kirk notes, "the commercial tone of early Canadian park policy resulted largely from the influencial [*sic*] role played by the railway companies in developing the first national parks," "The Development of National Park Policy," 60. While this statement is correct and reflects contemporary opinion about the nature of national parks at the turn of the century, it perhaps overstates the case for commercial attraction of parks. The original park act for Banff was modelled on the legislation creating Yellowstone National Park in 1872. And, as Roderick Nash has argued in "The American Cult of the Primitive," in Nash, ed., *The American Environment*, national parks were early endowed with a moral dimension.

62 Canada, "An Act Respecting Forest Reserves and Parks," *Laws and Statutes*, 1–2 Geo. 5 c.. 10.

63 Harkin, "*The History and Meaning of the National Parks of Canada*," 7.

64 Ibid., 81.

65 "Report of the Commissioner of Dominion Parks, in Department of the Interior, *Annual Report* (1914): 4.

66 Ibid. In the same report Harkin presented the case for national parks in a way typical of progressive thought. "Within recent years there has been a movement, particularly in Europe and the United States, generally referred to as the 'Recreation Movement'. It has had much to do with the wonderful progress that has been made with respect to supervised playgrounds for children, but its field is

much greater than that, and concerns play for adults as well. It had its origin through recognition of the fact that modern social and industrial conditions are resulting in a suppression or a perversion of the 'play spirit' and that this spells danger for the nation as well as the individual," ibid., 5.

67 Lothian, *History of Canada's National Parks*, vol. II: 12.

68 "Report of the Commissioner of Dominion Parks," in Department of the Interior, *Annual Report* (1913): 10.

69 Ibid. (1914): 11.

70 Ibid.

71 Fort Howe, in Saint John, New Brunswick, was created a national park under the authority of section 18 of the Dominion Forest Reserves Act in 1914. Although this was the original national historic park, it was never very successful and was declassified and handed over to the municipality in 1930.

72 *Spectator*, (Annapolis Royal) 22 Feb. 1918.

73 For example, the British National Trust for the Preservation of Places of Historic Interest or Natural Beauty and the American Scenic and Historic Preservation Society. The annual report for 1914 of this second organization is reprinted in Nash, ed., *The American Environment*, 78ff.

74 Williamson to H. Piers, curator, Halifax Museum, 3 Sept. 1913, NA, RG 84, vol. 1189, file HS 6 (Maritimes General), pt 1.

75 Harkin to J.G. Mitchell, 1 Mar. 1919, ibid.

76 Clarance M. Warner, president, Ontario Historical Society, to William J. Roche, minister of the Interior, 3 July 1914, ibid., NA, RG 84, vol. 1329, file HS 9 (Historic Sites in Western Ontario), pt 1.

77 Harkin to Mitchell, 1 March 1919, ibid., NA, RG 84, vol. 1189, file HS 6, pt 1.

CHAPTER TWO

1 For example, Harkin's memorandum advising the minister on the need for an advisory board read in part: "I would suggest that an honorary committee be appointed ... to advise the Department in the matter of preserving those sites which pre-eminently possess Dominion-wide interest." J.B. Harkin to J.G. Mitchell, 1 Mar. 1919, NA, RG 84, vol. 1189, file HS 6, pt 1.

2 Harkin to Mitchell, 21 June 1920, NA, RG 84, vol. 1329, file HS 9, pt 1.

3 One section of the proposed legislation reads: "If the Governor in Council is of the opinion that any site, situate [*sic*] partly or wholly within the lands of any private person, which is deemed to be of national historic interest and importance is in danger of

destruction, damage or removal in whole or in part, he may make an order to be known as a 'preservation order', placing such site for the time being under the protection and control of the minister." 9 Feb. 1920, NA, RG 84, Vol. 1434, file HS 12, pt 1.

4 Canada. House of Commons, *Debates*, 1920, 3283.

5 Voorhis, *Historical Account of the Territorial Expansion of the Dominion of Canada.*

6 According to the *Dominion Government Telephone Directory* for 1923, the parks branch consisted of the following components: commissioner, deputy commissioner, engineers, historic sites, accountant, law and lands clerk, forest protection, director of publicity, migratory birds and park animals, town planning.

7 The activities of the embryo wildlife division in this period are recounted in Foster, *Working for Wildlife,* chapter 7.

8 Harkin's annual report for 1919 noted: "Throughout the year the work in connection with the parks service has been planned and carried out primarily with a view to bringing into Canada a revenue of millions of dollars from foreign tourist traffic." "Report of the commissioner," in Department of the Interior, *Annual Report* (1919): 3.

9 "It is on his skillful array of facts that Colonel Cruikshank always depends for his force. Impartial and judicial, he may sometimes fail to interest but he never fails to satisfy." Anon. review of *Battle of Queenston Heights,"* 67–8.

10 Cited by McConnell, "E.A. Cruikshank, His Life and Work," 125.

11 Coyne, ed. and trans., "Exploration of the Great Lakes 1669–1670."

12 Cited by Stevenson, "James H. Coyne," 147.

13 Audet, "Benjamin Sulte," 338.

14 Morgan, ed., *Canadian Men and Women of the Time,* 2d. ed., 1076.

15 Morgan, ed., *Canadian Men and Women of the Time,* 986.

16 Morgan, ed., *Canadian Men and Women of the Time,* 2d. ed., 809.

17 Macleod, "Our Man in the Maritimes," 90.

18 W.C. Milner to J.D. Hazen, 14 Feb. 1914, Borden Papers, NA, MG 26, vol. 163. H 1(c).

19 A.G. Doughty to R.L. Borden, 23 Feb. 1914, ibid.

20 Captain Harry J. Knight to F.H.H. Williamson, 30 Aug. 1920, NA, RG 84, vol. 1048, file FB 2, vol. 1.

21 Milner to Williamson, 30 Aug. 1920, NA, vol. 1089, HS 6, pt 2.

22 Milner to Borden, 29 Oct. 1920, Meighen Papers, NA, MG 26, I, vol. 29.

23 W.O. Raymond to Williamson, 2 Aug. 1920, NA, RG 84, vol. 1089, file HS 6, pt 1.

24 Milner lived to a great old age and continued to beset the board and the parks branch into the 1930s. A later board member related the

following story: "He has just lost two sisters at a very advanced age. One of them left instructions that her dear brother Bill was not to be notified of her death, as she did not wish him to attend her funeral." J.C. Webster to Harkin, 8 Dec. 1930, NA, RG 84, vol. 1189, HS 6, pt 6.

25 Responding to a suggestion in Parliament that the heritage program be transferred to the National Battlefields Commission, Charles Stewart, minister of the Interior, said: "I have not very much information with respect to the Quebec Battlefields Commission, but I do know that the Historic Sites Commission have been sitting for some time, and taking into consideration the merits of the various historic sites throughout Canada with a view to preserving them and keeping them in repair." Canada, House of Commons, *Debates*, 1923, 3406.

26 At a meeting of 24 November 1919 attended by Cruikshank, Sulte, Harkin, and Williamson, it was resolved "that the Secretary be instructed to visit Louisbourg, Fort Edward, Fort Cumberland and other such sites as may be directed and report upon the necessary repairs which should be undertaken." Minutes, HSMBC.

27 Milner to W.W. Cory, 24 Dec. 1919, NA, RG 84, vol. 1189, HS 6, pt 1. He also expressed his concern at the next full meeting of the board where it was recorded that "Mr. Milner asked for the privilege of having the powers and status of the Board defined." Minutes, HSMBC, 18 May 1920.

28 Harkin to Cory, 2 Jan. 1920, NA, RG 84, vol. 1189, HS 6, pt 1.

29 Milner to Borden, 14 June 1920, Borden Papers, NA, MG 26, H, vol. 163.

30 Harkin to G. Buskard, private secretary to the prime minister, 8 Dec. 1920, NA, MG 26, I, vol. 29.

31 Williamson, circular letter to historical societies, 13 Sept. 1920, NA, RG 84, vol. 1329, file HS 9, pt 1.

32 Minutes, HSMBC, 18 May 1920.

33 American historian Henry F. May has described a similar sense of mission existing in the cultural élite in the United States in the decades before the First World War. He coined the term "custodians of culture" to describe the collective attitudes of this group. *The End of American Innocence*, 30.

34 James H. Coyne to E.A. Cruikshank, 13 Aug. 1920, NA, RG 84, vol. 1329, HS 9, pt 1.

35 E.S. Caswell to Harkin, 21 Nov. 1928, NA, RG 84, vol. 1330, file HS 9, pt 6 (Historic Sites in Western Ontario).

36 Coyne to Harkin, 2 Mar. 1922, NA, RG 84, vol. 1329, HS 9, pt 2.

37 NA, RG 84, acc. 83–84/280, vol. 1015, file HS 9–4, pt 2.

38 Cruikshank to Harkin, 26 Jan. 1921, ibid., pt 1.

39 Raymond to Harkin, 18 Jan. 1921, ibid.
40 Coyne to Cruikshank, 13 Aug. 1920, NA, RG 84, vol. 1329, HS 9, pt 1.
41 Minutes, HSMBC, 21 May 1921.
42 Coyne to Webster, 1 Feb. 1929, Webster Collection, drawer no. 1, item 41, archives section, New Brunswick Museum.
43 Cruikshank to Williamson, 13 Oct. 1920, NA, RG 84, 83–84/280, vol. 1305, HS 8, pt 1.
44 N.d., NA, RG 84, vol. 1234, HS 7, pt 1.
45 Benjamin Sulte to Harkin, 14 Nov. 1919, NA, RG 84, vol. 1238, HS 7–1, pt 1.
46 Sulte to Williamson, 19 June 1920, ibid.
47 *Documents Relative to the Colonial History of the State of New York*, vol. 3 (Albany, 1853), 800ff.
48 NA, MG 1, series C 11 A, vol. 11, 558ff.
49 Sulte, *Mélanges historiques*, vol. 1: 133.
50 Lighthall *Old Measures: Collected Verse*, 39.
51 Victor Morin to Williamson, 3 Nov. 1920, NA, RG 84, vol. 1237, file HS 7, pt 1.
52 These sites had particular relevance to the Action française which flourished under the leadership of Lionel Groulx in the 1920s. Susan Mann Trofimenkoff has noted the contemporary significance of Dollard's sacrifice at the Long Sault to this group. "Dollard belonged less to history than to the Action Française. For it, Dollard portrayed all the traits that the Action Française advocated for young French Canadians: he was religious, strong, brave, dominant, patriotic, and self-sacrificing." *Action française: French-Canadian Nationalism in the Twenties*, 43. The same attributes could be applied to Valrennes at Laprairie.
53 NA, RG 84, vol. 1240, file HS 7–4–1, pt 1.
54 Minutes, HSMBC, 21 May 1921.
55 Cruikshank to Harkin, 11 Jan. 1920, NA RG 85, vol. 1240, HS 7–4–1.
56 Denison, "The United Empire Loyalists and their Influence upon the History of this Continent," in RSC, *Proceedings and Transactions*, XXXIII.
57 Minutes, HSMBC, 18 May 1920.
58 Williamson to Harkin, Nov. 1919, NA, RG 84, vol. 1189, HS 6, pt 1.
59 Minutes, HSMBC, 21 May 1921.
60 R.W. Tufts to Harkin, 24 Apr. 1922, NA, RG 84, vol. 1048, FB 2 (Fort Beauséjour), pt 1.
61 Sulte to Williamson, 14 Feb. 1920, NA, RG 84, vol. 1061, FC 2 Fort Chambly, pt 1.
62 Thibodeau, "La conservation du Fort Chambly, 1850–1940," 34.
63 Harkin to Cory, 13 Nov. 1920, NA, RG 84, vol. 1061, FC 2, pt 1; Order-in-Council, 10 Jan. 1921.

64 Pinard to Harkin, 13 Dec. 1920, NA, RG 84, vol. 1061, FC 2, pt 1.
65 Thibodeau, *La conservation du Fort Chambly*, 36.
66 Report of Committee of Privy Council, 18 May 1921, NA, RG 84, vol. 1084, file FLE 2 (Fort Lennox), pt 2.
67 Pinard to Harkin, 30 Sept. 1921, ibid.
68 Cruikshank to Harkin, 22 July 1921, ibid.
69 Minutes, HSMBC, 18 May 1920.
70 Eugene Fiset, deputy minister of Militia and Defence to Cory, deputy minister of the Interior, 14 Mar. 1922, NA, RG 84, vol. 1312, HS 8–12, pt 1.
71 Pinard to Cruikshank, 11 Dec. 1920, NA, RG 84, vol. 1329, HS 9, pt 1.
72 Ibid.
73 Cruikshank to Williamson, 28 June 1920, ibid., FPW 2, pt 1.
74 Harkin to Cory, 25 June 1920, ibid.
75 CHA, *Report* (1924): 101.
76 Memo to file by Pinard, 30 Nov. 1920, NA, RG 84, vol. 1374, HS 10, pt 1.
77 An implacable critic of the parks branch, Thomas L. Church (Lib., Toronto North), proposed the extension of the mandate of the National Battlefields Commission to include Ontario sites. Canada, House of Commons, *Debates*, 1922, 119.
78 Frederick Godsal to F.W. Howay, 14 Apr. 1923, NA, RG 84, vol. 1374, HS 10, pt 2.
79 "Our Historic Sites' officers have standing orders to prepare a memorandum looking forward to a reorganization [of the board]. However, through Williamson's sickness and Pinard's repeated absences the memorandum has not yet been prepared but I have reminded them several times of it and no doubt will receive it at an early date." Harkin to Cruikshank, 19 Nov. 1920, NA, RG 84, vol. 1329, HS 9, pt 1.

CHAPTER THREE

1 J.B. Harkin to J.C. Webster, 5 Jan. 1923, NA, RG 84, vol. 1173, HS General, pt 4; Harkin to J.P. Edwards, 8 Jan. 1923.
2 Minutes, HSMBC, 19 May 1925.
3 Ibid., May 1926.
4 Ibid.
5 W.W. Cory to Harkin, 13 Jan. 1930, NA, RG 84, vol. 1173, HS General, pt 7. "The Minister has discussed a number of times with the Minister of National Defence a scheme for the establishment of a Commission to take care of such historic sites as St. Louisbourg [*sic*], Halifax, Kingston and the Quebec walls." In reply, Pinard advised Harkin (A.A.P. to J.B.H., 16 Jan. 1930) that, "as the branch has already successfully dealt with the restoration, preservation, mainte-

nance, and working of a large number of military and other structures, ... I cannot see why it could not handle the new problems quite as readily and efficiently as any other body. Of course, this would necessitate additional money and assistance."

6 Harkin to Cory, 18 Dec. 1926, ibid., HS General pt 5.

7 Memorandum from Harkin to R.A. Gibson, acting deputy minister of Interior re preservation in Halifax, Québec, and Kingston, 31 May 1929, NA, RG 84, vol. 1312, HS 8–12 (Fort Henry), pt 1.

8 Harkin to A.A. Pinard, 26 Dec. 1928, NA, RG 84, vol. 1173, HS General, pt 5.

9 "An Act Respecting National Historic Sites of Canada," 1924, NA, RG 84, vol. 1434, HS 12 (legislation), pt 1.

10 W. Crowe to Hon. R.E. Harris, 18 May 1923, NA, RG 84, vol. 1095, FLO 2 (Louisbourg establishment), pt 4.

11 A major project for Adams at this time was "a skillfully landscaped design for the Jasper National Park headquarters town." Michael Simpson, "Thomas Adams in Canada," 8. I am grateful to Professor John Taylor of Carleton University for bringing this reference to my attention.

12 "Memo on town planning," ca. 1921, NA, RG 84, vol. 1172, HS General pt 3.

13 T.R. Adams to Harkin, 11 June 1923, NA, RG 84, acc. 83–84/280, vol. 340, FLO 2, pt 4.

14 Harkin to Crowe, 22 Jan. 1926, ibid., FLO 2, pt 6.

15 Harkin to Cory, 22 Jan. 1923, NA, RG 84, vol. 1189, HS 6, pt 1. "It is suspected that Mr. Fortier supplies these statements. He has been very persistent in his efforts to obtain a free hand in regard to the park and has complained bitterly against every action of the Department in checking up his work."

16 Harkin to L.M. Fortier, 25 July 1930, NA, RG 84, vol. 1041, FA 2, pt 4.

17 Fortier to Harkin, 18 Dec. 1922, NA, RG 84, vol. 1805, PR 2 (Port Royal establishment), pt 1.

18 Edwards to Harkin, 24 Feb. 1923, ibid.

19 Harkin to C. Whitman, 29 June 1927, Ibid., PR 2, pt 2.

20 Jefferys, "The Reconstruction of the Port Royal Habitation," 370.

21 Canada, House of Commons, *Debates*, 1929, 3647.

22 Harkin to Cory, 26 Nov. 1928, NA, RG 84, vol. 1805.

23 Harkin to Gibson, assistant deputy minister of the Interior, 28 Feb. 1929, ibid.

24 Minutes, HSMBC, 26 May 1923.

25 G. Lanctôt to Pinard, 22 May 1922, NA, RG 84, vol. 1085, FLE 2 (Fort Lennox establishment) pt 3.

26 Harkin to C.A. Papineau, 9 April 1924, ibid., FLE 2, pt 7.

27 Canadian National Parks Branch, "Some Historic and Pre-historic Sites of Canada," in CHA, *Report* (1925): 88.
28 Ibid. (1927): 107.
29 Minutes, HSMBC, 20 May 1929.
30 Denys Nelson, *Fort Langley*, 29.
31 Canadian National Parks Branch, "Some Historic and Pre-historic Sites of Canada," CHA, Report (1927): 107.
32 Memorandum to file by G.W. Bryan, 1 Dec. 1923, NA, RG 84, vol. 1189, HS (Maritimes General).
33 J.H. Coyne to Harkin, 15 Dec. 1923, NA, RG 84, vol. 1329, file HS 9, pt 3. "For example," he wrote, "how is the line to be drawn in respect to the following:
 1. Pioneer settlements, churches, schools, highways, posts, towns, villages
 2. Industrial establishments and organizations
 3 Agricultural development
 4. Social and literary organizations."
34 Minutes, HSMBC, 25 May 1923.
35 Minutes, HSMBC, 25 May 1925. The board's reference in these selections appears to have been George Johnson, *Alphabet of First Things in Canada*, Pinard to M. McCormack, 23 July 1924, NA, RG 84, vol. 1234, file HS 7, pt 3.
36 F.W. Howay to Harkin, 8 March 1927, NA, RG 84, vol. 1375, file HS 10, pt 4. Three sites proposed for designation that year were: First Coal Mine in Alberta, Barkerville, British Columbia and Yukon gold discovery.
37 E.A. Cruikshank to Harkin, 10 Oct. 1924, NA, RG 84, vol. 1305, file HS 8, pt 2. "The Construction of the various canals under government control were indisputedly important events in the industrial history of Canada, which it is thought should be commemorated at some convenient future date."
38 Webster to Harkin, 28 July 1928, NA, RG 84, vol. 1189, HS 6, pt 5.
39 Minutes, HSMBC, 4 June 1924.
40 Ibid.
41 F.H.H. Williamson to Gibson, 6 April 1937, NA, RG 84, vol. 1306, HS 8, pt 5.
42 Canada, House of Commons, *Debates*, 1923, 3406.
43 Webster, *Those Crowded Years*, 17.
44 Webster to Charles Stewart, 22 Dec. 1922, NA, RG 84, vol. 1189, HS 6, pt 1.
45 Pinard to Cruikshank, 15 Sept. 1922, ibid., HS 6, pt 2.
46 Webster to Harkin, 19 July 1926, NA, RG 84 vol. 1190, HS 6, pt 5. The board, however, decided that a token site should be designated on

the island and "moved that a sub-committee composed of Dr. Webster and Major Edwards be nominated to take up the question of a suitable memorial to commemorate an historical event in Prince Edward Island." Minutes, HSMBC, 26 May 1923.

47 Webster, *The Distressed Maritimes*, 13.
48 Webster to Howay, 4 Dec. 1942, F.W. Howay Papers, box 7, UBCL.
49 Crowe to Webster, 1927, Webster Collection, drawer no. 1, New Brunswick Museum.
50 Crowe to Harkin, 31 Mar. 1928, NA, RG 84, vol. 1173, HS General, pt 6.
51 Minutes, HSMBC, 17 May 1928.
52 Ibid., 17 May 1929.
53 Ibid., 20 May 1929.
54 The particular emphasis of his commemorations was described by Webster in his memoirs: "It was my aim to induce the Dominion Government to carry out measures which would result in bringing the attention of our people the points of historic interest in the entire Chignecto area, the marking of its historic sites, the preservation of the Fort and the establishment of a Historical museum – lines of development which had been successfully carried out at Fort Anne in Annapolis Royal and at Ticonderoga on Lake Champlain." *Those Crowded Years*, 20.
55 Cruikshank to Howay, 3 May 1929, box 2, Howay Papers.
56 Webster to Harkin, 11 Oct. 1924, NA, RG 84, vol. 1095, FLO 2, pt 6.
57 Edwards to Harkin, 15 Aug. 1923, ibid.
58 Harkin to Crowe, 9 Mar. 1926, ibid.
59 Canada, House of Commons, *Debates*, 1928, 2625.
60 Ibid., 244.
61 Webster, *Those Crowded Years*, 23.
62 Memorandum to file, 7 May 1928, NA, RG 84, vol. 1096, FLO 2, pt 8. Twelve thousand dollars of this appropriation, however, was needed for the acquisition of real estate.
63 Webster to Sen. J.S. McLennan, 21 Dec. 1928, McLennan Collection, Louisbourg National Historic Park. My thanks to A.J.B. Johnston, staff historian at Louisbourg, for bringing this document to my attention.
64 Canada, House of Commons, *Debates*, 1929, 3647.
65 In forming his cabinet in 1925, "King was dubious about the capacities of Dr. King from British Columbia and Charles Stewart from Alberta but saw no suitable alternatives among the few Liberals elected from these provinces." Neatby, *William Lyon Mackenzie King*, vol. II: 172.
66 "There is one point which I think should be stressed in this connec-

tion and that is that the benefit to follow from the preservation of any of these old sites are enjoyed primarily by the city in which they are located and secondly by the Province. I therefore am strongly of the opinion that in regard to Halifax, Quebec and Kingston some sort of a policy should be worked out which would involve co-operation on the part of the city, the province and the Dominion." Harkin to Gibson, 31 May 1929, NA, vol. 1312, HS 8–12, pt 1.

67 Harkin to Gibson, 11 Apr. 1929, NA, RG 84, vol. 1096, FLO 2, pt 10.

68 Ruskin, *Seven Lamps of Architecture*, 179.

69 Harkin to J.G. Mitchell, 31 May 1920, Borden Papers, NA, RG 26, H vol. 163.

70 In 1922 when McLennan wrote Harkin asking that steps be taken to preserve Louisbourg, he made it clear just what he envisioned when he referred to the description of the original buildings in his book. "It has full plans and will indicate to you how complete a restoration of the site could be made if funds and intelligence were available." McLennan to Harkin, 1 Dec. 1922, NA, RG 84, vol. 1095, FLO 2, pt 1.

71 Hosmer, *Presence of the Past*, 92.

72 Webster to Harkin, 20 Nov. 1923, NA, RG 84, vol. 1095, FLO 2, pt 6.

73 Crowe to Webster, 27 Apr. 1929, Webster Collection, drawer no. 1, archives section.

74 Harkin to Gibson, 11 Apr. 1929, FLO 2, pt 10.

75 Crowe wrote to Webster, 27 April 1929: "Our work for Lbg. is not even begun and we must consider what we are to do at Ottawa about it. I wish you would think over a good line or approach – my own view is we ought to have Stuart and Ralston together and talk the situation with them very frankly – or should we have a talk with Harkin and try to win him to a special Commission – and then go to the minister." Webster Collection, drawer no. 1.

76 Cruikshank opposed this arrangement. In a letter to Harkin, 3 April 1931, he said: "the proposal to constitute a 'Local Advisory Committee' to advise a sub-committee of this Board, appointed to advise the Board, which was appointed to advise the Minister, appears to me to be extraordinary and inadvisable and certain to cause complications." NA, RG 84,m vol. 10995, FLO 2, pt 13.

77 Webster wrote to Harkin, 15 Oct. 1928, "this work must now go ahead. The people demand it, and it is admitted by all that it should have been done years ago. The Prime Minister, Mr. Bennett, and Col. Ralston have promised their support, so that their [*sic*] should be no difficulty unless it should come from your department." Ibid., FLO 2, pt 9.

78 Crowe to Harkin, 1 Aug. 1928, FLO 2, pt 8.

79 "I wish to say that there has been general disappointment expressed by many of the visitors there the past summer – I am referring to the American motor car visitors of whom we have had a large number this year." Crowe to Harkin, 17 Oct. 1929, ibid., FLO 2, pt 11.

80 "The Preservation of Fort Louisbourg," n.d., ibid., FLO 2, pt 17.

81 Ibid.

82 W.L. Ormond to Harkin, 16 May 1923, NA, RG 84, vol. 1048, FB 2, (Fort Beauséjour establishment), pt 2.

83 Webster to Harkin, 29 Sept. 1924, ibid., pt 3.

84 Harkin to Webster, 21 June 1926, ibid., pt 4.

85 "Memorandum Re: Improvements at Fort Cumberland Historic Site," 22 Nov. 1926, ibid.

86 Webster to Harkin, 7 Sept. 1926, ibid. Webster closed his letter with a threat: "You may consider me out of it entirely unless a thorough scientific method is adopted."

87 Memorandum to file by Pinard, 22 Feb. 1928, NA, RG 84, vol. 1189, HS 6, pt 5.

88 Webster, *Those Crowded Years*, 26.

89 Victor Morin to Harkin, 20 Mar. 1924, NA, RG 84, vol. 1173, HS General, pt 4.

90 Minutes, HSMBC, 4 June 1924.

91 Howay to Cruikshank, 28 June 1935, Howay Papers, box 9.

92 Sage, "Frederic William Howay," 448.

93 Howay, "The Earliest Pages of the History of British Columbia," 16.

94 Howay to H.H. Stevens, 15 Aug. 1932, Howay Papers, box 8.

95 Howay was for a time president of a local cultural society called The Fellowship of Arts. "Each year – that is each season, October to April – we have a program of study. This year we are taking the time of Shakespeare, as a sort of complement to last year, which was the time of Queen Elizabeth." Howay to Cruikshank, 20 Feb. 1934, Howay Papers, box 8.

96 Minutes, HSMBC, 16 May 1928.

97 He submitted his list to Harkin, 5 Nov. 1923, saying, "bearing in mind that I am representing the region west of the Great Lakes I have striven to give a fair proportion to each province." The historic sites, in order of priority were: Yale, Nootka Sound, Batoche, Duck Lake, Fort Livingstone, Fort Macleod, Fort Edmonton, Fort Langley, Fort George, and Frog Lake.Howay to Harkin, 5 Nov. 1923, NA, RG 84, vol. 1376, HS 10 (Western Canada General), pt 2.

98 For an account of this controversy see my "Some Early Problems of the Historic Sites and Monuments Board."

CHAPTER FOUR

1 Treasury Board Minutes, 7 Aug. 1931, NA, RG 32, C 2, vol. 466 (A.A. Pinard).
2 F.W. Howay to E.A. Cruikshank, 24 Sept. 1937, Howay Papers, box 9, UBCL.
3 In a memorandum to the assistant deputy minister of the Interior, Harkin noted that "inasmuch as our historic sites' appropriation has been reduced by the sum of $12,000 for the present year, it will not be possible to carry out any of the items of work included in my memorandum of the 15th June ... other possibly, than to affix the tablet for the First Paper Mill to the cairn at St. Andrews East, PQ." J.B. Harkin to R.A. Gibson, 18 July 1933, NA, RG 84, acc. 83–84/280, vol. 904, HS General, pt 8.
4 "Report of the Commissioner of National Parks." in Department of the Interior, *Annual Report* (1934): 122.
5 J.C. Webster to Harkin, 28 July 1928, NA, RG 84, vol. 1189, HS 6, pt 5.
6 Harkin to Webster, 19 Oct. 1932, ibid., HS 6, pt 7.
7 W. Crowe to F.H.H. Williamson, 18 Mar. 1931, ibid.
8 Howay to Cruikshank, 30 Oct. 1935, Howay Papers, box 9.
9 Harvey, "History and its Uses in Pre-Confederation Nova Scotia," 12.
10 D.C. Harvey to Webster, 12 Feb. 1935, Webster Collection, drawer no. 1, archives section, New Brunswick Museum.
11 Minutes, HSMBC, 30 May 1935.
12 Diary entry, 13 Jan. 1931, Landon Papers, box 4210, file 42, regional collection, D.B. Weldon library, University of Western Ontario.
13 Fred Landon and Jesse E. Middleton, *The Province of Ontario: A History, 1645–1927*, 4 vols. (Toronto: Dominion Publishers 1927).
14 Armstrong, "Fred Landon, 1880–1969," 2–4.
15 Landon, *Western Ontario and the American Frontier*, 280–1.
16 Fred Landon to Harkin, 3 Mar. 1935, NA, RG 84, vol. 1330, file HS 9, pt 8. Howay to Harkin, 18 Mar. 1935; Minutes, HSMBC, 30 May 1935.
17 Howay, for example, took a personal dislike to Walter N. Sage who emerged to share the leadership of the British Columbia historical movement in the 1930s. He particularly disliked his pretensions to the title doctor and referred to his insufferable manner. Howay to T.C. Elliot, 3 May 1937 and 23 July 1937, Howay Papers, box 9.
18 A history of material culture studies in the United States has noted a similar phenomenon occurring in the country in a much earlier period with the advent of university trained historians. "In their celebration of the textual data of American history, the scientific historians chose to ignore completely the artifactual evidence of the

national past. As they became concentrated in the universities, the professional historians of the late-nineteenth and early twentieth centuries abandoned their former affiliations with various local and state historical associations, where an assortment of material culture studies were being done by collectors, curators, and amateur researchers. Soon museum historians, avocational historians, and gentlemen scholars like James Ford Rhodes came to feel, largely because they lacked the credential of a doctorate in history and did not teach history in a college or university, that they were excluded from the emerging national forum of scholarly activity and cameraderie that became the American academic historical profession." Schlereth, ed., "Material Cultural Studies in America" 13–14.

19 In the introduction Landon noted "several of the earlier chapters were read by the late Brigadier-General E.A. Cruikshank whose extensive knowledge of Ontario's early history made his counsel of particular value." Landon, *Western Ontario and the American Frontier*, xv.

20 E.-F. Surveyer to Webster, 30 Jan. 1930, Webster Collection, drawer no. 1.

21 21 Jan. 1933, ibid.

22 Memorandum to file by G.W. Bryan, 4 Nov. 1930, NA, RG 84, vol. 1173, HS General, pt 7.

23 Minutes, HSMBC, 30 May 1934.

24 Ibid., 27 May 1933.

25 Ibid., 19 May 1925.

26 Memorandum to file by A.A. Pinard, 23 May 1929, NA, RG 84, vol. 1189, file HS 6, pt 6.

27 Howay to Cruikshank, 30 Oct. 1935, Howay Papers, box 9.

28 Minutes, HSMBC, 29 May 1934.

29 "L'association des architectes de la Province de Québec créait un club qui organise des concours, donne des cours et encourage le relevé des anciens édifices de la province." Anon., *Canadian Architect and Builder* XVIII, 208 (avril 1909): 49. I am grateful to Janet Wright of the architectural history division of Parks Canada for bringing this reference to my attention.

30 In the introduction Traquair explained: "This is a book about buildings, their form, construction and decoration, about the traditions which led to those forms, about the materials and the techniques employed." Traquair, *The Old Architecture of Quebec*, xiii.

31 Surveyer, in Royal Architectural Institute of Canada, *Journal* III, 3 (May-July 1926): 117.

32 Way, *Ontario's Niagara Parks*, 253.

33 Way, however, maintained that it was the Niagara Parks Commission's interest "in the early preservation of national history, more

than considerations of scenic beauty, which underlay the acquisition of park areas at Queenston, Fort Erie, Lundy's Lane and Niagara-on-the-Lake." Ibid., 241.

34 Way, "Old Fort Henry," 164.

35 Minutes, HSMBC, 27 May 1933.

36 Ibid., 29 May 1934.

37 Harvey to Harkin, 15 Mar. 1934, NA, RG 84, vol. 1174, HS General, pt 8.

38 Cited in McConnell, "Cruikshank, his Life and Work", 92.

CHAPTER FIVE

1 Struthers, *No Fault of their Own*, 82.

2 "Report of the Commissioner of National Parks," in Department of the Interior, *Annual Report* (1932): 91.

3 W. Crowe to J.C. Webster, 8 Nov. 1833, Webster Collection, drawer no. 1, New Brunswick Museum.

4 Crowe to Senator McLennan, Apr. 1933, ibid.

5 "Report of the Commissioner of National Parks," in Department of the Interior, *Annual Report* (1935): 101.

6 Ibid. (1936): 93.

7 R.A. Gibson to J.B. Harkin, 18 June 1935, NA, RG 84, vol. 1805, PR 2, pt 2; Harkin to Gibson, 11 June 1935, ibid.

8 Thibodeau, "La conservation du Fort Chambly," 41.

9 H.H. Rowatt to Harkin, 26 Nov. 1931, NA, RG 84, vol. 1130, FPW 2, pt. 4.

10 Undated memorandum, "Reconstruction of Fort Prince of Wales," by George Kydd (resident engineer, Department of Railways and Canals, Churchill), ibid., FPW 2, pt 5. Kydd's men used a dragline, a Holt tractor, a concrete mixer, and wagons to facilitate their work.

11 National Parks Branch, Department of the Interior, "Preserving Canada's Historic Past," in CHA, *Report* (1936): 119.

12 According to W.F. Lothian, who served in the parks bureau through the 1930s, Bennett regularly telephoned Harkin to request his resignation. Personal communication, 14 Aug. 1985.

13 Lothian, *History of Canada's National Parks*, vol. II: 18.

14 At this meeting Crerar said: "A country which forgets its history, which loses sight of the actors on the stage of the historic past, it seems to me is a country which is fast approaching the point where decay sets in. It is perhaps with this point of view in mind that I have been so very much impressed with the value of the Board's undertakings." Minutes, HSMBC, 24 May 1944.

15 Williamson was the only official formally attached to the board, but

it is evident that Cromarty acted in the capacity of assistant secretary until his formal elevation to secretary in 1943. Howay noted at this time, "Mr. Cromarty has been so long associated with our work in a clerical capacity that I had almost forgotten his advancement," Minutes, HSMBC, 19 May 1943.

16 Adell, "Upper Hot Springs Bathhouse."

17 Harkin, *History and Meaning of the National Parks of Canada*, 13.

18 Adell, "Upper Hot Springs Bathhouse."

19 In the cabinet discussions preceding the 1936 budget, for example, Crerar argued for tax exemptions to stimulate the mining industry. Neatby, *William Lyon Mackenzie King*, vol. 3: 163.

20 Auditor General, *Annual Report* (1936), (1950).

21 Neatby, *William Lyon Mackenzie King*, vol. 3: 250–5.

22 Neatby argues that this budget represented a significant break in previous government thinking and marked the introduction of Keynesian fiscal policy. Ibid., 256–7.

23 F.H.H. Williamson to R.A. Gibson, 11 June 1938, NA, RG 84, vol. 1174, HS General, pt 9.

24 F.W. Howay to E.A. Cruikshank, 18 Dec. 1938, Howay Papers, box 9, UBCL.

25 Jefferys, "The Reconstruction of the Port Royal *Habitation*," 370–1. Schmeisser, "Port Royal Habitation, 1928–1938."

26 Jefferys suggest that the sudden decision stemmed from the fact that the key piece of real estate came on the market. But the parks service had already established the price of the property in 1935 and could have presumably expropriated the property if it had been guaranteed a commitment of funds. K.D. Harris to Harkin, 24 June 1935, NA, RG 84, vol. 1805, PR 2, pt 2.

27 Minutes, HSMBC, 28 May 1934.

28 Cruikshank to Harkin, 17 Oct. 1936, NA, RG 84, vol. 1805, file PR 2, pt 2.

29 UNESCO, "International Charter for the Conservation and Restoration of Monuments and Sites," Venice 1966. Article 15 reads in part: "Ruins must be maintained and measures necessary for the permanent conservation and protection of architectural features and of objects discovered must be taken. Furthermore, every means must be taken to facilitate the understanding of the monument and to reveal it without ever distorting its meaning. All reconstruction work should however be ruled out *a priori*. Only anastylosis, that is to say, the reassembling of existing but dismembered parts can be permitted."

30 "The sources of information regarding the Habitation are the engravings in the works of Champlain and Lescarbot, the text of these

contemporary writers, and some letters in the Jesuit *Relations*. To these should be added the illuminating comments of Dr. Ganong on the maps of Champlain, based on his intimate local knowledge; and the researches of Mrs. Harriette Taber Richardson, an American lady, who since 1923 has spent her summers in the neighbourhood, and who has made an intensive study of the source documents, and a continued and detailed exploration of the site and the surrounding country." Jefferys, "Reconstruction of the Port Royal Habitation," 370.

31 Schmeisser, "Port Royal Habitation," 25.
32 Department of Mines and Resources, *Annual Report* (1944): 87.
33 E.-F. Surveyer to Webster, 15 Oct. 1940, Webster Collection, drawer no. 1, item 47, archives section.
34 Schmeisser, "Port Royal Habitation," 35.
35 The plaque has since been removed. Schmeisser, ibid., 36.
36 D.C. Harvey to Webster, 25 Sept. 1946, Webster Collection, drawer no. 1, item 47, archives section.
37 John Rick to D.C. Smith, 15 Sept. 1967. Cited in Schmeisser, "Port Royal Habitation."
38 J.A. Smart to Gibson, 23 Sept. 1937, NA, RG 84, vol. 1931, file SWL 2, pt 2.
39 Williamson to Surveyer, 21 Dec. 1937, ibid. Surveyer forgot about the telephone call and, when the purchase of the property was announced publicly in December, he sent a sharp note to Williamson claiming that he had not been consulted. Surveyer to Williamson, 20 Dec. 1937, ibid.
40 F. Todd to C.W. Jackson, 6 Aug. 1938, ibid.
41 G.W. Bryan to Williamson, 14 Sept. 1938, ibid.
42 Dec. 1938, ibid., SWL 2, pt 3; Todd to Jackson, 22 Dec. 1938, ibid.
43 "The sum of $4,000 is provided in the Supplementary Estimates for repairs and improvements to the birthplace of Sir Wilfrid Laurier at St. Lin, together with the purchase of suitable exhibits which would enable us to convert the building into an historic museum." Williamson to Gibson, 28 July 1938, NA, RG 84, vol. 1235, HS 7, pt 9.
44 Smart to Williamson, 8 July 1940, NA, RG 84, vol. 1931, SWL 2, pt 3.
45 "La famille Laurier a résidé à cet endroit mais la maison qu'on y retrouve présentement n'est peut-être pas la sienne; ce que l'on tente de démontrer en disant qu'elle n'est plus à son emplacement d'origine ou bien que c'est pas celle-ci." Rainville, "Le parc historique national de la maison Sir Wilfrid Laurier," 119.
46 Cited in McConnell, "Cruikshank, His Life and Work," 92.
47 Carter-Edwards, "Fort Malden," vol. 1: 294.
48 Ibid., 296.

49 Ibid., 297.
50 W.D. Cromarty to J.E. Spero, 25 Nov. 1940, NA, RG 84, vol. 1174, HS General, pt 11.

CHAPTER SIX

1 F.W. Howay to J.C. Webster, 2 Feb. 1943, Webster Collection, drawer no. 1, New Brunswick Museum.
2 E.A. Cruikshank to J.C. Coyne, 10 Mar. 1939, Coyne Papers, regional collection, D.B. Welson Library, University of Western Ontario.
3 Webster to Howay, 10 Feb. 1943, Webster Collection.
4 D.C. Harvey to Webster, 19 Nov. 1938, Webster Collection.
5 12 Oct. 1934, ibid.
6 Thomas, "Morden Heaton Long," 298.
7 Minutes, HSMBC, 31 May 1939. Cruikshank and Howay were both commemorated by secondary markers following their deaths.
8 Harvey to Webster, 16 Jan. 1939, Webster Collection.
9 E.-F. Surveyer to F.H.H. Williamson, 26 Feb. 1940, NA, RG 84, vol. 1174, HS General, pt 11; Surveyer to J.B. Harkin, 5 Feb. 1936, NA, RG 84, vol. 1375, HS 10, pt 8.
10 Referring to a letter from Jacob Livinson, chairman of the citizenship committee of the (Montréal) City Improvement League, to commemorate the first synagogue in Canada, Surveyer stated frankly, "I am not particularly interested in the commemoration of Jewish activities." Surveyer to W.D. Cromarty, 20 June 1945, NA, RG 84, vol. 1236, HS 7, pt 10.
11 Howay to Williamson, 24 June 1937, ibid.
12 A. D'Eschambault to Williamson, 4 Jan. 1939, ibid. Cromarty to Father I.J. Lesiuk, 19 May 1945, ibid., vol. B76, MS 10, pt 9.
13 Minutes, HSMBC, 1943.
14 W.N. Sage to Cromarty, 27 Sept. 1948, NA, RG 84, vol. 1376, HS 10.
15 Sen. Pierre Casgrain to Cromarty, 8 Nov. 1940, NA, RG 84, vol. 1236, HS 7, pt 10.
16 Howay to Williamson, 11 Feb. 1941, ibid.
17 Webster to Williamson, 10 Feb. 1941, ibid.
18 Harvey to Williamson, 11 Feb. 1941, ibid.
19 Reginald Harris to Williamson, 12 Jan. 1938, NA, RG 84, vol. 1190, HS 6, pt 8.
20 La Maison Fargues. NA, RG 84, vol. 1235, HS 7, pt 9.
21 Williamson to Paul Martin, MP, 29 May 1940, NA, RG 84, vol. 1331, HS 9, pt 10.
22 Cruikshank to Williamson, 9 Nov. 1938, NA, RG 84, vol. 1306, HS 8,

pt 5. The board deferred the request to preserve the blockhouse. Minutes, HSMBC 29 May 1939.

23 Minutes, HSMBC, 19 May 1943.
24 Ibid.
25 Ibid.
26 Royal Commission, *Report*, xi.
27 Ibid., xii.
28 Massey, *What's Past is Prologue*, 213. Bissell, *The Young Vincent Massey*, 215–21.
29 Another important issue was university funding. Pickersgill, *My Years with Louis St. Laurent*, 140.
30 Hilda Neatby, *So Much To Do, So Little Time*, ed. Michael Hayden, 27.
31 Royal Commission, *Report*, 124.
32 Ibid.,126; 127.
33 Ibid., 347.
34 Ibid.
35 Ibid.,348.
36 Ibid.,350.
37 Harvey to C.G. Childe, 17 Sept. 1951, NA, RG 84, vol. 1183, HS 2.
38 Fred Landon to Childe, 10 March 1952, ibid.
39 Sage to Childe, 9 Nov. 1951, ibid.
40 Minutes, HSMBC, 27–30 May 1952.
41 Landon to Childe, 30 Sept. 1952, NA, RG 84, vol. 1183, HS 2.

CHAPTER SEVEN

1 "Secretary's Report," in Minutes, HSMBC, Nov. 1962.
2 Pickersgill, *My Years with Louis St. Laurent*, 206.
3 A bilateral agreement setting out rights between Canada and the United States over tapping the river for irrigation and hydroelectric dams.
4 Lindo, "Interpretation in Science and Technology, Museums. The Alexander Graham Bell National Historic Park: A Case Study," 255.
5 "Report on Restoration of the Halifax Citadel (Fort George) prepared at the Request of the Royal Commission on National Development in the Arts, Letters and Sciences," NA, RG 84, vol. 1162, HC 8R, pt 1.
6 E.A. Côté, to J.A. Hutchison, 5 Sept. 1956, ibid., HC 28, vol. 8, pt 2.
7 Gordon Scott to J.D. Herbert, 3 May 1961, NA, RG 84, Vol. 1441, HS 28, vol. 1.
8 J.R.B. Coleman to Côté, 24 Nov. 1960, NA, RG 84, vol. 1162, HC 28, vol. 9.

9 Côté to the minister, 23 Mar. 1959, NA, RG 84, vol. 1245, HS 7–12, vol. 1.
10 Côté to Parks Branch, 12 Jan. 1959, NA, RG 84, vol. 1303, HS 7–172.
11 Coleman to Hutchison, 18 July 1955, NA, RG 84, vol. 1441, HS 28. Vol.
12 Hutchison to A.J.H. Richardson, 19 Aug. 1955, NA, RG 84, vol. 11011, FLO 28, vol. 4, pt 1. Hutchison to deputy minister, 30 July 1956, ibid., vol. 5, pt 3.
13 Côté to minister, 26 Feb. 1959, NA, RG 84, vol. 1102, FLO 28, vol. 6, pt 5.
14 Lower, *My First Seventy-Five Years*, 354.
15 Fred Landon to Maurice Lamontagne, 14 July 1954, NA, RG 84, vol. 1176, HS 1, Vol. 15, pt 1.
16 "Minutes of the Meeting of the Committee on Criteria of the Historic Sites and Monuments Board," held in room 305, Norlite Building, Ottawa, in Minutes, HSMBC, 1, 2 Apr. 1962.
17 Coleman to Côté, 8 June 1966, NA, RG 84, Vol. 1181, HS 1, (Thematic).
18 Way, "Upper Canada Village," 218–33.
19 Bureau, "Historic Monuments in Urban Centers."
20 "Preliminary Statement," Confederation of Tomorrow Conference, Toronto, November 1967, in *Canadian Federalism: Myth or Reality*, ed., Meekison, 364.

CHAPTER EIGHT

1 "The changes in public attitudes towards conservation came as an almost sudden reaction to the 'biggest is best' era of the 1950's and 1960's when communities were disintegrated, ploughed up and concreted over at an unprecedented rate." RSC, *Preserving the Canadian Heritage*, 39.
2 Minutes, HSMBC, June, 1954.
3 Ibid., Dec. 1955.
4 A.J.H. Richardson to Fred Landon, 21 Mar. 1956, NA, RG 84, vol 1443, HS 56, vol. 3.
5 Minutes, HSMBC, June, 1957.
6 A.R.M. Lower to Walter Dinsdale, 17 June 1961, NA, RG 84, vol. 1182, HS 1 – Lower A.
7 Landon to Richardson, 15 June 1956, NA, RG 84, vol. 1443, HS 56, vol. 4.
8 "Report of the Committee Established to Study the Preservation of Historic Buildings." Minutes, HSMBC, Nov. 1958.
9 E.A. Côté to W.H. Cranston, 14 Dec. 1960, NA, RG 84, vol. 1444, HS 56, vol. 7.
10 Richardson to John Bland, 6 Apr. 1956, NA, RG 84, vol. 1443, HS 56, vol. 4.
11 Minutes, HSMBC, May 1959.

12 Heritage Canada.
13 Landon to Jean Lesage, 20 Aug. 1956, NA, RG 84, vol. 1175, HS General, vol. 13, pt. 1.
14 Alvin Hamilton to Patricia Clarke, 5 Aug. 1960, NA, RG 84, vol. 1444, HS 56, vol. 7.
15 Minutes, HSMBC, May 1958.
16 Ibid., Nov. 1958.
17 Côté to Coleman, 20 Apr. 1959, NA, RG 84, vol. 1301, H 57–155, vol. 1.
18 "Over and above the Historic importance of the property, however, there is the fact that this attempt at restoration is the first undertaken by this Department ... It promises to be also, one of the largest actual restoration jobs to be accomplished in Canada by any group or agency." Coleman to Alan Jarvis, 25 May 1959, NA, RG 84, vol. 1301, 457–155, vol 3.
19 Minutes, HSMBC, May 1953; HS, May 1957.
20 A. D'Eschambault to Richardson, 14 Sept. 1957, NA, RG 84, vol. 1420, HS 10–154, vol. 1.
21 J.D. Herbert to Coleman, 23 Aug. 1959, ibid., HS 10–154, vol. 2, pt 2.
22 Clifford P. Wilson to Herbert, 5 Oct. 1959, ibid.
23 Hamilton to A. D'Eschambault, 25 Feb. 1960, ibid.
24 Herbert to Richardson, 18 Jan. 1962, ibid., HS 10–154, vol. 2, pt 1.
25 H.J. Johnston to T.C. Fenton, n.d. [ca. July 1962], ibid.
26 Côté to minister, 20 Mar. 1967, NA, RG 84, vol. 1175, HS General, Vol. 16, pt 1.
27 Côté to branch, 18 Oct. 1960, NA, RG 84, vol. 1444, HS 56, vol. 7.
28 "Marking and Preservation of Historic Sites: Activity Summary of the Federal Government," 14 Oct. 1960, NA, RG 84, vol. 1175, HS General, vol. 14, pt 1.

CHAPTER NINE

1 J.A. Hutchison to deputy mininster, 5 Jan. 1956, NA, RG 84, vol. 1176, 45 1, vol. 16, pt1.
2 Minutes, HSMBC, 12 Dec. 1955.
3 Ibid.
4 Bothwell, Drummond, English, *Canada Since 1945,* 200.
5 R.G. Robertson to G.E. Steele, 27 Feb. 1961, NA, RG 84, vol. 1425, HS 10–205, vol. 1.
6 Minutes, HSMBC, May 1959. The minister addressed the board personally in November, when he said: "The Yukon, particularly areas related to the Gold Rush Era, have tremendous tourist potential. It is essential that a true historical picture be extricated from this image of romance and legend. The Board is asked to give this mat-

ter thought. Since this is a federally-controlled region consideration could be given to adapting municipal regulations to the needs of historical preservation." Minutes, HSMBC, Nov. 1959.

7 Ibid.

8 Ibid.

9 Dawson was largely a ghost town. In rehabilitating one central building, it was natural that the scope would be expanded to include other interesting old buildings and attractions, such as the Robert Service Cabin, to justify the attention paid to the Palace Grand.

10 T.C. Fenton to chief engineer, 14 July 1960, NA, RG 84, vol. 1425, HS 10–205, vol. 1.

11 J.R.B. Coleman to E.A. Coleman, 22 nov. 1960, NA, RG 84, vol. 1425, HS 10–205, vol. 1.

12 Ibid.

13 Pete Schonenbach to J.D. Herbert, 9 Dec. 1960, ibid.

14 J.R.B. Coleman to deputy minister, 7 Feb. 1961, ibid.

15 "Authority to Enter into Contract," 10 Feb. 1961, ibid.

16 Coleman to E.A. Côté, 19 June 1961, ibid.

17 Memorandum to file by Côté, 28 Feb. 1961, ibid.

18 Ibid.

19 R.G. Robertson to G.E. Steele, 27 Feb. 1961, ibid.

20 Côté to Coleman, 14 Feb. 1961, ibid.

21 Stewart, "Recycling Used Boom Towns," 15.

22 Associated Boards of Trade of Cape Breton, "Twenty-Two Point Development Program," 6 Aug. 1958, NA, RG 84, vol. 1102, FLO–28, vol. 6, pt 5.

23 Côté to minister, 26 Feb. 1959, ibid.

24 H.A. Johnson to A.D. Perry, 16 Dec. 1958, ibid.

25 K. McLennan to Herbert, 23 Sept. 1960, ibid., FLO 28, vol. 6, pt 3.

26 Herbert, "Report on Trip to Sydney, Baddeck and Louisbourg, March 11–13," 15 Mar. 1960, ibid., FLO 28, vol. 6, pt 2.

27 Royal Commission on Coal, *Report*, 53.

28 "Planning Considerations: The Historic Parks – Maritimes," report reviewed at Guidance Committee on Planning 17–25 May 1960, NA, RG 84, vol. 1441, HS 28, vol. 1.

29 Ibid.

30 Ibid.

31 Norman P. Robinson to G.L. Scott, 12 May 1960, NA, RG 84, vol. 1102, FLO 28, vol. 6, pt 1.

32 Robertson to Côté, 14 Oct. 1960, NA, RG 84, vol. 1102, FLO 28, vol. 6, pt 1.

33 Draft memo to cabinet, 24 Feb. 1961, ibid., vol. 8.

34 Coleman to deputy minister, 27 Apr. 1960, ibid., vol. 6, pt 1.

35 Ibid.

36 Scott to D.P. Nigra, 4 Oct. 1963, NA, RG 84, vol. 1184, HS 3.
37 Ronald Way, "Historical Restoration."
38 Way, "Recommendations Concerning the Louisbourg Restoration Project."
39 Ibid.
40 Ibid.
41 Ibid.
42 Herbert to Coleman, 26 Sept. 1961, NA, RG 84, vol. 1102, FLO 28, vol. 8, pt 3.
43 Coleman to Perry, 13 Oct. 1961, ibid., pt 2.
44 Scott to A. Reeve, 27 Feb. 1962, ibid. vol. 9, pt 1.
45 Scott to Coleman, 26 Feb. 1962, ibid., pt 2.
46 Minutes, HSMBC, May 1962.
47 Ronald L. Way to Coleman, 8 Oct. 1962, NA, RG 84, vol. 1103, FLO 28, vol. 11.
48 Coleman to Perry, 10 Oct. 1962, ibid.
49 Way to Coleman, 30 July 1963, ibid., vol. 13, pt 3.
50 F.J. Thorpe, "Some Thoughts on Policy," Oct. 1963, ibid., pt 2.
51 J.H.Rick to Herbert, 30 Aug. 1963, ibid., vo. 1441, HS 28, vol. 2.
52 Coleman to Côté, 1 Mar. 1963, ibid.
53 Rick to Herbert, 30 Aug. 1963, ibid.
54 "Minutes of Staff Meeting, Louisbourg 29 and 30 Oct. 1963, NA, RG 84, vol. 1103, FLO 28, vol. 13, pt 2.
55 Ibid.
56 Ibid.
57 Ibid.
58 Palardy, "Guide to Chateau St. Louis." MacLean, "Historical Research at Louisbourg."
59 Côté to minister, 20 mar. 1967, NA, RG 84, vol. 1175, H.S. General, vol. 16, pt1.
60 "The Historical Program of the National and Historical Parks Branch of Canada: A Statement of Purpose, Goals and Methods," second draft," n.d. NA, RG 84, vol. 1441, HS 28, vol. 1.
61 "Minutes of staff meeting Louisbourg, 29 and 30 Oct. 1963," NA, RG 84, vol. 1103, FLO 28, vol. 13, pt 2.
62 "Fortress of Louisbourg Restoration Section," ibid., FLO 28, vol. 14, pt 2.
63 John I. Nicol to senior assistant deputy minister, 22 Sept. 1966, ibid., pt 1.
64 George Galt, "Making History," 132.
65 In 1978, on its way to visit this site, a contingent of the Historic Sites and Monuments Board was killed in a tragic air crash.
66 Department of the Environment, *Parks Canada Policy*, 69.

Bibliography

MANUSCRIPT SOURCES

British Columbia

Vancouver

University of British Columbia Library. Special Collections Division. F.W. Howay Papers.

New Brunswick

Saint John

New Brunswick Museum. Archives Section. Webster Manuscript Collection.

Ontario

London

D.B. Weldon Library. University of Western Ontario. Regional Collection:

James Coyne Papers
Fred Landon Papers

Ottawa

Department of the Environment. Canadian Parks Service. Historic Sites and Monuments Board of Canada. Minutes.

National Archives of Canada. Federal Archives Division:

National Battlefields Commission Records
Parks Canada Records
Public Service Commission Records

National Archives of Canada. Manuscript Division:

R.L. Borden Papers
E.A. Cruikshank Papers
G.T. Denison III Papers
Earl Grey Papers
J.B. Harkin Papers
Wilfrid Laurier Papers
Arthur Meighen Papers

GOVERNMENT DOCUMENTS

Canada

Auditor General. *Annual Report* 1920–50.

Canadian Parks Service. "Guide to Chateau St Louis" by Jean Palardy. Unpublished manuscript on file.

Department of the Environment. *Parks Canada Policy.* Ottawa: Department of Supply and Services 1982.

Department of the Interior. *Annual Report* 1914–35.

Department of Mines and Resources. *Annual Report* 1936–49.

Department of Resources and Development. *Annual Report* 1950.

Dominion Government Telephone Directory 1923–50.

House of Commons. *Debates* 1911–40.

House of Commons. *Journal* 1907–40.

Parks Canada. National Historic Parks and Sites Branch. "The Archaeological Excavations at the Southwold Earthworks" by David Smith. Manuscript Report Series no. 314: 1935.

– "Miscellaneous Research Paper" by William Russell. Manuscript Report Series no. 216: 1975–7.

– "La conservation du Fort Chambly, 1850–1940" par Pierre Thibodeau. Travail inédit no. 377: 1979.

– "The Creation and Development of Fort Beauséjour National Historic Park" by Barbara Schmeisser, Manuscript on file 1979.

– "Fort Malden: A Structural Narrative History 1796–1976" by Dennis Carter-Edwards. Vol. 1. Manuscript Report Series no. 40: 1980.

– Direction des lieux et des parcs historiques nationaux. "Le parc historique national de la maison Sir Wilfrid Laurier" par Alain Rainville. Rapports sur microfiche no. 62: 1982.

– Federal Heritage Buildings Review Office. "Upper Hot Springs Bathhouse Banff National Park, Alberta" by Jaqueline Adell. Building Report 84–5: 1984.

Royal Commission on Coal. *Report* 1960.

Royal Commission on National Development in the Arts, Letters and Sciences. *Report* 1951.

Royal Commission on National Development in the Arts, Letters and Sciences. *Royal Commission Studies* 1951.

Senate. *Debates* 1895–1940.

Ontario

Minister of Education. "Archaeological Report." *Annual Report* 1891.

International

International Council of Monuments and Sites. International Charter for the Conservation and Restoration of Monuments and Sites (Venice Charter) 1966.

BOOKS AND ARTICLES

Anon. "Review of Battle of Queenston Heights." *Review of Historical Publications Relating to Canada* 9: 67–8.

Armstrong, Fred. "Fred Landon 1880–1969." *Ontario History* 62: 1–4.

Arthur, E.A. "The Early Architecture of the Province of Ontario." Royal Architectural Institute of Canada *Journal* V (1928).

– *The Early Buildings of Ontario.* Toronto: University of Toronto Press 1938.

Audet, Francis J. "Benjamin Sulte." *Le bulletin des recherches historiques* 32, 6 (June 1926): 337–43.

Berger, Carl. "The True North Strong and Free." In *Nationalism in Canada,* edited by Peter Russell. Toronto: McGraw-Hill 1966.

– *The Sense of Power; Studies in the Ideas of Canadian Imperialism, 1867–1914.* Toronto: University of Toronto Press 1970.

– *The Writing of Canadian History: Aspects of English-Canadian Historical Writing, 1900–1970.* Toronto: Oxford University Press 1976.

Bernatchez, Ginette. "La Société littéraire et historique du Québec (Literary and Historical Society of Québec) 1824–1890." *Revue d'histoire de l'Amérique française* 34: 179–92.

Bothwell, Robert, Ian Drummond, and John English. *Canada Since 1945: Power, Politics, and Provincialism.* Toronto: University of Toronto Press 1981.

Brown, Robert, and Ramsay Cook. *Canada 1896–1921: A Nation Transformed.* Toronto: McClelland and Stewart 1974.

Brunet, Michel. *La présence anglaise et les Canadiens: études sur l'histoire et la pensée des deux Canadas.* Montréal: Beauchemin 1958.

Buffie, Erna. "The Massey Report and the Intellectuals: Tory Cultural

Nationalism in Ontario in the 1950s." Master's thesis, University of Manitoba 1982.

Bureau, Pierre. "Historic Monuments in Urban Centres." Paper presented at the 11th Annual Conference on Historic Resources, 1971, Québec City.

Calnan, David Mikel. "Businessmen, Forestry and the Gospel of Efficiency, the Canadian Conservation Commission: 1909–1921." Master's thesis, University of Western Ontario 1975.

Careless, J.M.S. "Limited Identities in Canada." *Canadian Historical Review* 50: 1–10.

– "The Review Reviewed or Fifty Years with the Beaver Patrol." *Canadian Historical Review* 51: 48–71.

Carr, E.H. *What is History?* London: Penguin Books 1964.

Cassirer, Ernst. *An Essay on Man: An Introduction to a Philosophy of Human Culture.* New Haven: Yale University Press 1944.

Chadwick, George F. *The Park and the Town: Public Landscape in the Nineteenth and Twentieth Centuries.* London: Architectural Press 1966.

Charlevoix, F.X. *History and General Description of New France.* New York: J.G. Shea 1902.

Champlain, Samuel de. *The Works of Samuel de Champlain.* Edited by H.P. Biggar. 6 vols. Toronto: Champlain Society 1922–36.

Commission des monuments historiques de la province de Québec. *Vieux manoirs, vieilles maisons.* Québec: Imprimeur du roi 1927.

Coyne, James H. *The Country of the Neutrals: From Champlain to Talbot.* St Thomas, Ont.: Times Printing 1895.

Cruikshank, E.A. *The Story of Butler's Rangers and the Settlement of the Niagara Area.* Niagara Falls, Ont.: Lundy's Lane Historical Society 1893.

– *Documentary History of the Campaign on the Niagara Frontier in 1812–14.* Welland, Ont.: Tribune 1896.

– ed. and trans. "Exploration of the Great Lakes 1669–1670 by Dollier de Casson and de Brehant de Galinée." *Ontario Historical Society Papers and Record,* IV (1903): 1–89.

Fawcett, Jane, ed. *The Future of the Past, Attitudes to Conservation, 1174–1974.* London: Thames and Hudson 1976.

Fergusson, C.B. "Daniel Cobb Harvey." *Canadian Historical Review* 47: 399–400.

Fitch, James Marston. *Historic Preservation: Curatorial Management of the Built World.* New York: McGraw-Hill 1982.

Foster, Janet. *Working for Wildlife, the Beginning of Preservation in Canada.* Toronto: University of Toronto Press 1978.

Galt, George. "Making History." *Saturday Night* 102, 1: 130–4.

Geist, Christopher. "Historic Sites and Monuments as Icons." In *Icons of America,* edited by Ray B. Browne and Marshall Fishwick. Bowling Green, Ohio: Popular Press 1988.

Gillis, Robert Peter. "The Ottawa Lumber Barons and the Conservation Movement, 1880–1914." *Journal of Canadian Studies* 9: 14–30

Granatstein, J.L. *The Ottawa Men: The Civil Service Mandarins, 1935–1957.* Toronto: Oxford University Press 1982.

Guitard, Michelle. *The Militia of the Battle of the Chateauguay: A Social History.* Ottawa: Parks Canada 1983.

Hallett, Mary Elizabeth. "The 4th Earl Grey as Governor General of Canada 1904–1911." PhD diss., University of London 1969.

Hardy, René. "L'ultramontanisme de Laflèche: genèse et postulats d'une idéologie." *Idéologies au Canada français, 1850–1900,* edited by Fernand Dumont et al. Québec: Les presses de l'université Laval 1971.

Harkin, J.B. *The History and Meaning of the National Parks of Canada, Extracts from the Papers of the Late Jas. B. Harkin, First Commissioner of the National Parks of Canada.* Compiled by Mabel B. Williams. Saskatoon: H.R. Larson Pub. Co. 1957.

Harvey D.C. "History and its Uses in Pre-Confederation Nova Scotia." In Canadian Historical Association. *Report* 1938: 5–16.

Harvey, John. *Conservation of Buildings.* Toronto: University of Toronto Press 1972.

Hays, Samuel P. *The Response to Industrialism: 1884–1914.* Chicago: University of Chicago Press 1957.

– Conservation and the Gospel of Efficiency; The Progressive Conservation Movement, 1890–1920. Cambridge, Mass.: Harvard University Press 1959.

Hewison, Robert. *The Heritage Industry: Britain in a Climate of Decline.* London: Methuen 1987.

Hitsman, J. McKay. *The Incredible War of 1812.* Toronto: University of Toronto Press 1965.

HLA. *Appeal for Membership* 1908.

– *Annual Report* 1915.

Hodgetts, J.E. *The Canadian Public Service: A Physiology of Government, 1867–1970.* Toronto: University of Toronto Press 1973.

Hopkins, J. Castell, ed. *Canadian Annual Review of Public Affairs.* Toronto: Annual Review Publishing Co. 1908.

Hosmer, Charles B. *Presence of the Past, A History of the Preservation Movement in the United States before Williamsburg.* New York: G.P. Putnam's Sons 1965.

– *Preservation Comes of Age: From Williamsburg to the National Trust, 1926–1949.* 2 vols. Charlottesville: University Press of Virginia 1981.

Howay, F.W. "The Earliest Pages of the History of British Columbia." British Columbia Historical Association. *First Annual Report* 1923: 16–22.

Huth, Hans. "The Evolution of Preservationism in Europe." *Journal of the American Society of Architectural Historians* 1: 5–9.

Ise, John. *Our National Park Policy, A Critical History.* Baltimore: Johns Hopkins Press 1961.

Jeffereys, Charles W. "The Reconstruction of the Port Royal *Habitation* of 1605–13." *Canadian Historical Review* 20: 369–77.

Johnston, A.J.B. "Preserving History: The Commemoration of 18th Century Louisbourg, 1895–1940." *Acadiensis* 12: 53–80.

Killan, Gerald. *Preserving Ontario's Heritage: A History of the Ontario Historical Society.* Ottawa: Ontario Historical Society 1976.

– *David Boyle: From Artisan to Archaeologist.* Toronto: University of Toronto Press 1983.

Kohn, Hans. *The Idea of Nationalism: A Study in its Origins and Background.* New York: Macmillan 1944.

Landon, Fred. *Western Ontario and the American Frontier.* Toronto: Ryerson Press 1941.

LeMoine, J.M. *Maple Leaves; A Budget of Legendary, Historical, Critical and Sporting Intelligence.* Québec: Hunter Rose 1863.

Lighthall, W.D. *An Account of the Battle of Chateauguay.* Montréal: W. Drysdale and Co. 1889.

– *Montreal after 250 Years.* Montréal: F.E. Grafton and Sons 1892.

– *Old Measures: Collected Verse.* Montréal: A.T. Chapman 1922.

Lindo, P. Richard. "Interpretation in Science and Technology Museums: The Alexander Graham Bell National Historic Park, A Case Study." In *Critical Issues in the History of Canadian Science, Technology and Medicine/Problèmes cruciaux de l'histoire de la science, de la technologie et de la médecine au Canada,* edited by Richard A. Jarrell and Arnold E. Roos. Thornhill, Ont.: HSTC Publications 1983.

Long, M.H. "The Historic Sites and Monuments Board of Canada." In Canadian Historical Association. *Annual Report* 1954: 1–11.

Lothian, W.F. *A History of Canada's National Parks.* 4 vols. Ottawa: Parks Canada 1976–81.

Lower, A.R.M. *My First Seventy-Five Years.* Toronto: Macmillan 1967.

Mackintosh, Barry. *The Historic Sites Survey and National Historic Landmarks Program: A History.* Washington, DC: Department of the Interior 1985.

Maclean, Terry. "Historical Research at Louisbourg: A Case Study in Museum Research and Development." In *Cape Breton at 200: Historical Essays in Honour of the Island's Bicentennial, 1785–1985,* edited by Ken Donovan. Sydney, NS: University College of Cape Breton Press 1985.

Macleod, Donald. "Our Man in the Maritimes: 'Down East' with the Public Archives of Canada," 1872–1932. *Archivaria* 17: 86–105.

Massey, Vincent. *What's Past is Prologue.* Toronto: Macmillan 1963.

May, Henry F. *The End of American Innocence; A Study of the First Years of our own Time, 1912–1917.* New York: Alfred A. Knopf 1959.

McConnell, David. "E.A. Cruikshank, His Life and Work." Master's thesis, University of Toronto 1965.

Meekison, J. Peter, ed. *Canadian Federalism: Myth or Reality.* Toronto: Methuen 1968,.

Monet, Jacques. *The Last Cannon Shot; A Study of French-Canadian Nationalism, 1837–1850.* Toronto: University of Toronto Press 1969.

Monière, Dennis. *Le développement des idéologies au Québec: des origines à nos jours.* Montréal: Éditions Québec/Amérique 1977.

Morgan, Henry James. *Canadian Men and Women of the Time.* Toronto: William Briggs 1898.

– *Canadian Men and Women of the Time.* 2d ed. Toronto: William Briggs 1912.

Murphy, Achille. "D'embellissements de la ville de Québec proposés par Lord Dufferin en 1875." *Journal of Canadian Art History* 1: 18–29.

Nash, Roderick. *The American Environment; Readings in the History of Conservation.* Reading, Mass.: Addison-Wesley Pub. Co. 1968.

– *The Nervous Generation: American Thought 1917–1930.* New York: Rand McNally 1970.

Neatby, H. Blair. *William Lyon Mackenzie King.* Vol. II, *The Lonely Heights, 1924–1932.* Toronto: University of Toronto Press 1963.

– *William Lyon Mackenzie King.* Vol. III, *The Prism of Unity, 1932–1939.* Toronto: University of Toronto Press 1976.

Nelles, H.V. *The Politics of Development; Forests, Mines and Hydro-Electric Power in Ontario, 1849–1941.* Toronto: Macmillan 1974.

Nelson, Denys. *Fort Langley, 1827–1927, A Century of Settlement in the Valley of the Lower Fraser River.* Vancouver: Evans Hastings 1927.

Nelson, J.G., ed. *Canadian Parks in Perspective.* Montréal: Harvest House 1970.

O'Callaghan, E.B., ed. *Documents Relative to the Colonial History of the State of New York.* 15 vols. Albany: Weed, Parsons 1853–87.

Parkman, Francis. *Count Frontenac and New France Under Louis XIV.* Toronto: G.N. Morang 1899.

– *Montcalm and Wolfe, France and England in North America.* 2 vols. Toronto: G.N. Morang 1899.

Peers, Frank W. *The Politics of Canadian Broadcasting, 1920–1951.* Toronto: University of Toronto Press 1969.

Pickersgill, J.W. *My Years with Louis St. Laurent: A Political Memoir.* Toronto: University of Toronto Press 1975.

Roy, Pierre-Georges. *Les monuments commémoratifs de la province de Québec.* 2 vols. Québec: King's Printer 1923.

Royal Society of Canada. *Preserving the Canadian Heritage.* Ottawa: Royal Society of Canada 1975.

Ruskin, John. *The Seven Lamps of Architecture.* London: Smith, Elder & Co. 1849.

Russel, Loris. *The National Museum of Canada 1910–1960.* Ottawa: Queen's Printer 1961.

Sage, W.N. "Frederic William Howay (1867–1943) FRSC, FRHS." *Canadian Historical Review* 24: 448.

Schlereth, Thomas J. *Artifacts and the American Past.* Nashville: American Association for State and Local History 1980.

– "Material Culture Studies in America, 1876–1976." In *Material Culture Studies in America,* edited by Thomas J. Schlereth. Nashville: The American Association for State and Local History 1982.

Schmeisser, Barbara. "Port Royal Habitation, 1928–1939: A Case Study in the Preservation Movement." Paper presented at the annual meeting of the Canadian Historical Association, May 1985, Montréal.

Simpson, Michael. "Thomas Adams in Canada, 1914–1930." *Urban History Review* 11: 1–15.

Stacey, C.P. "The War of 1812 in Canadian History." *Ontario History* 50, no. 3 (1958): 153.

– *A Date With History: Memoirs of a Canadian Historian.* Ottawa: Deneau 1983.

Stevenson, Hugh A. "James H. Coyne, An Early Contribution to Canadian Historical Scholarship." *Ontario History* 54: 25–42.

– "James H. Coyne: His Life and Contributions to Canadian History." Master's thesis, University of Western Ontario 1960.

Stewart, Richard. "Recycling Used Boom Towns: Dawson and Tourism." Paper presented to Boomtown Conference, University of Victoria, Victoria, BC, August 1987.

Struthers, James. *No Fault of their Own: Unemployment and the Canadian Welfare State, 1914–1941.* Toronto: University of Toronto Press 1983.

Sulte, Benjamin. *Histoire des Canadiens-français, 1608–1880.* 8 vols. Montréal: Société de publication historique du Canada 1884.

– Mélanges historiques. 21 vols. Montréal: DuCharme 1918–34.

Surveyer, E.-F. "Address by the Honourable Judge E. Fabre Surveyer, as representative of McGill University at the Annual Banquet of the Royal Architectural Institute of Canada." Royal Architectural Institute of Canada *Journal* 3, 3 (June 1926): 117.

– *The First Parliamentary Elections in Lower Canada.* Montréal: L. Carrier 1927.

Taylor, C.J. "Historic Sites." *Canadian Encyclopedia.* vol. 2. Edmonton: Hurtig Publishers 1985.

– "National Historic Parks and Sites, 1880–1951: The Biography of a Federal Cultural Program," PhD diss., Carleton University, 1986.

– "Some Early Problems of the Historic Sites and Monuments Board of Canada." *Canadian Historical Review* 64: 3–24.

– "World Heritage Convention." *Canadian Encyclopedia.* 2d ed. vol. 4. Edmonton: Hurtig Publishers 1988.

Thomas, Lewis H. "Morden Heaton Long." *Canadian Historical Review* 46: 298.

Traquair, Ramsay. *The Old Architecture of Quebec.* Toronto: Macmillan 1947.

Trofimenkoff, Susan Mann. *Action française: French-Canadian Nationalism in the Twenties.* Toronto: University of Toronto Press 1975.

Turner, Robert D., and William E. Rees. "A Comparative Study of Parks Policy in Canada and the United States." *Nature Canada* 2: 31–6.

Underhill, Frank. "Notes on the Massey Report." In *In Search of Canadian Liberalism.* Toronto: Macmillan 1960.

Van Kirk, Sylvia M. "The Development of National Park Policy in Canada's Mountain National Parks, 1885 to 1930." Master's thesis, University of Alberta 1969.

Vipond, Mary. "National Consciousness in English-Speaking Canada in the 1920s: Seven Studies." PhD diss., University of Toronto 1974.

Voorhis, Ernest. *Historical Account of the Territorial Expansion of the Dominion of Canada from 1497 to 1920.* Ottawa: Department of the Interior 1922.

– *Historic Forts and Trading Posts of the French Regime and of the English Fur Trading Companies.* Ottawa: Department of the Interior 1930.

Wadland, John Henry. *Ernest Thompson Seton, Man in Nature in the Progressive Era 1880–1915.* New York: Arno Press 1978.

Wallace, W. Stewart. *The Macmillan Dictionary of Canadian Biography.* 3d. ed. Toronto: Macmillan 1963.

Way, Beryl. "Upper Canada Village." *Canadian Geographical Journal* 62: 218–33.

Way, Ronald W. "Old Fort Henry, the Citadel of Upper Canada." *Canadian Geographical Journal* 39: 148–69.

– *Ontario's Niagara Parks, A History.* Toronto: Niagara Parks Commission 1946.

Webster, John Clarence. *The Distressed Maritimes, A Study of Educational and Cultural Conditions in Canada.* Toronto: Ryerson Press 1926.

– *Those Crowded Years 1863–1944, An Octogenarian's Record of Work.* Shediac: privately published 1944.

Wilson, I.E. "Shortt and Doughty: The Cultural Role of the Public Archives of Canada." Master's thesis, Queen's University (Kingston) 1973.

– "Shortt and Doughty: The Cultural Role of the Public Archives of Canada, 1904–1035." *The Canadian Archivist/L'Archiviste canadien* 2: 4–25.

Wood, William. *Unique Quebec.* Québec: Literary and Historical Society of Québec 1924.

Young, Brian. *George-Étienne Cartier: Montréal Bourgeois.* Montréal: McGill-Queen's University Press 1981.

Index